The Ingredients

The Ingredients

A GUIDED TOUR OF THE ELEMENTS

Philip Ball

OXFORD

UNIVERSITY PRESS

OXFORD
UNIVERSITY PRESS

Great Clarendon Street, Oxford OX2 6DP

Oxford University Press is a department of the University of Oxford.
It furthers the University's objective of excellence in research, scholarship,
and education by publishing worldwide in

Oxford New York

Auckland Bangkok Buenos Aires Cape Town Chennai
Dar es Salaam Delhi Hong Kong Istanbul Karachi Kolkata
Kuala Lumpur Madrid Melbourne Mexico City Mumbai Nairobi
São Paulo Shanghai Singapore Taipei Tokyo Toronto

with an associated company in Berlin

Oxford is a registered trade mark of Oxford University Press
in the UK and in certain other countries

Published in the United States
by Oxford University Press Inc., New York

© Philip Ball 2002

British Library Cataloguing in Publication Data

Data available

Library of Congress Cataloging in Publication Data

Data available

ISBN 0–19–284100–9

1 3 5 7 9 10 8 6 4 2

Typeset in New Baskerville
by RefineCatch Limited, Bungay, Suffolk
Printed by T.J. International, Padstow, Cornwall

Preface

When I was asked to write an introduction to the elements as a companion volume to my book *Stories of the Invisible*, itself an introduction to molecules, I had mixed feelings. After all, in the earlier book I had been perhaps less than respectful towards the Periodic Table, that famous portrait of all the known chemical elements. Specifically, I had suggested that chemists cease to promote the notion that chemistry begins with this table, since a basic understanding of molecular science need embrace only a very limited selection of the hundred or more elements that the table now contains. No piano tutor would start by instructing a young pupil to play every note on the keyboard. Far better to show how just a few keys suffice for constructing a host of simple tunes. As music is about tunes, chords, and harmonies, not notes *per se*, so chemistry is about compounds and molecules, not elements.

But no one who is a chemist at heart can resist the elements, and that includes me. It includes Oliver Sacks too, who as a boy set about collecting the elements as most other boys collected stamps or coins. He wanted to own them all. In the 1940s it was not so hard to add to one's collection: Sacks could go to Griffin & Tatlock in Finchley, north London, and spend his pocket money on a lump of sodium, which he would then send fizzing over the surface of Highgate Ponds near his home. I envy him; the best I could do was to smuggle lumps of sulphur and bottles of mercury out of the school laboratory.

These elements were like precious stones or exquisite confectioneries. I wanted to touch and smell them, although prudence held me back from tasting. This tactile, sensual experience was made more poignant by the knowledge that these substances were pure, unalloyed, irreducible. They were the primal stuff of creation, sitting in my hand.

So I knew I would not be able to resist the lure of writing about the elements. But I began to see also that an introduction to the elements need not after all become a tour of the Periodic Table—which anyway others have conducted before me, and more skilfully or more exhaustively than I would be able to manage. The story of the elements is the story of our relationship with matter, something that predates any notion of the Periodic Table. Intimacy with matter does not depend on a detailed knowledge of silicon, phosphorus, and molybdenum; it flows from the pleasurable density of a silver ingot, the cool sweetness of water, the smoothness of polished jade. That is the source of the fundamental question: what is the world made from?

So there are 'elements' in this book that you will find in no Periodic Table: water and air, salt, subtle phlogiston. No matter that chemistry has now pulled them apart or banished them entirely; they are part of the table's legacy, and part of our pool of cultural symbols.

I am extremely grateful for the comments, advice, and materials I have received on various specific topics in this book from Al Ghiorso, Darleane Hoffmann, Scott Lehman,

Jens Nørskov, and Jim White. My thanks go also to Shelley Cox for her enthusiasm and faith in commissioning the book.

Philip Ball

London
March 2002

Contents

List of figures

Aristotle's Quartet

The Elements in Antiquity

In 1624 the French chemist Étienne de Clave was arrested for heresy. De Clave's inadmissible ideas did not concern the interpretation of holy scripture. Nor were they of a political nature. They did not even challenge the place of man in the universe, as Galileo was doing so boldly.

Étienne de Clave's heresy concerned the elements. He believed that all substances were composed of two elements—water and earth—and 'mixts' of these two with three other fundamental substances or 'principles': mercury, sulphur, and salt. It was not a new idea: the great French pharmacist Jean Béguin, who published *Tyrocinium chymicum* (*The Chemical Beginner*), one of the first chemistry textbooks, in 1610, maintained until his death a decade later that all matter had essentially those same five basic ingredients.

But want of originality did not help Étienne de Clave. His idea was heretical because it contradicted the system of elements propounded by the ancient Greeks and endorsed by Aristotle, their most influential philosopher. Aristotle took this scheme from his teacher Plato, who in turn owed it to

Empedocles, a philosopher who lived during Athens's Golden Age of Periclean democracy in the fifth century BC. According to Empedocles there were four elements: earth, air, fire, and water.

Shocked into cultural insecurity by the fall of Rome, the medieval West emerged from the trauma of the Dark Ages with a reverence for the scholars of antiquity that conflated their beliefs with the doctrines of Christianity. The word of Aristotle became imbued with God's authority, and to question it was tantamount to blasphemy. Not until the late seventeenth century did the discoveries of Galileo, Newton, and Descartes restore the Western world's ability to think for itself about how the universe was arranged.

Which is why the plan of Étienne de Clave and a handful of other French intellectuals to debate a non-Aristotelian theory of the elements at the house of Parisian nobleman François de Soucy in August 1624 was squashed by a parliamentary order, leading to the arrest of its ringleader.

The controversy was not really about science. The use of law and coercion to defend a theory was not so much an indication that the authorities cared deeply about the nature of the elements as a reflection of their wish to preserve the status quo. Like Galileo's trial before the Inquisition, this was not an argument about 'truth' but a struggle for power, a sign of the religious dogmatism of the Counter-Reformation.

Free of such constraints, the ancient Greeks themselves discussed the elements with far more latitude. The Aristotelian quartet was preceded by, and in fact coexisted with, several other elemental schemes. Indeed, in the sixteenth century the Swiss scholar Conrad Gesner showed that no fewer than eight systems of elements had been proposed between the times of Thales (the beginning of the sixth century BC) and Empedocles. The Condemnation of 1624 notwithstanding, this eventually made it harder to award any privileged status to Aristotle's quartet, and helped to open up again the question of what things are made from.

What are things made from? This is a short book, but the answer can be given even more concisely. Chemistry's Periodic Table lists all the known elements and, apart from the slowly growing bottom row of human-made elements, it is comprehensive. Here is the answer. These are the elements: not one, not four, not five, but about ninety-two that appear in nature.

What are things made from? The Periodic Table is one of the pinnacles of scientific achievement, but it does not quite do justice to that question. Set aside the fact that the atomic building blocks are actually more subtly varied than the table implies (as we shall see later). Forget for a moment that these atoms are not after all fundamental and immutable, but are themselves composites of other entities. Let us not worry for now that most people have never even heard of many of these elements, let alone have the vaguest notion of what they look and behave like. And make it a matter for discussion elsewhere that the atoms of the elements are more often than not joined into the unions called molecules, whose properties cannot be easily intuited from the nature of the elements themselves.* Even then, it is not enough to present the Periodic Table as if to say that Aristotle was wildly wrong about what things are made from and so was everyone else until the late eighteenth century. In asking after the elements, we can become informed about the nature of matter not just by today's answer (which is the right one), but by the way in which the problem has been broached in other times too. In response, we are best served not by a list but by an exploration of the enquiry.

What are things made from? We have become a society obsessed with questions about composition, and for good reason. Lead in petrol shows up in the snow fields of Antarctica; mercury poisons fish in South America. Radon from the earth poses health hazards in regions built on granite, and natural arsenic contaminates wells in Bangladesh. Calcium supplements combat bone-wasting diseases; iron alleviates anaemia. There are elements that we crave, and those we do our best to avoid.

The living world is, at first glance, hardly a rich dish of elements. Just four of them are endlessly permuted in the molecules of the body: carbon, nitrogen, oxygen, and hydrogen. Phosphorus is indispensable, not only in bone but in the DNA molecules that orchestrate life in all its forms. Sulphur is an important component of proteins, helping to hold them in their complex shapes. But beyond these key players is a host of others that life cannot do without. Many are

---

* Molecules are the topic of the companion volume to this book, *Stories of the Invisible* (Oxford: Oxford University Press, 2001).

metals: iron reddens our blood and helps it to transport oxygen to our cells, magnesium enables chlorophyll to capture the energy of sunlight at the foot of the food pyramid, sodium and potassium carry the electrical impulses of our nerves. Of all the natural elements, eleven can be considered the basic constituents of life, and perhaps fifteen others are essential trace elements, needed by almost all living organisms in small quantities. ('Toxic' arsenic and 'sterilizing' bromine are among them, showing that there is no easy division of elements into 'good' and 'bad'.)

The uneven distribution of elements across the face of the earth has shaped history—stimulating trade and encouraging exploration and cultural exchange, but also promoting exploitation, war, and imperialism. Southern Africa has paid dearly for its gold and the elemental carbon of its diamonds. Many rare but technologically important elements, such as tantalum and uranium, continue to be mined from poor regions of the world under conditions (and for reasons) that some consider pernicious and hazardous.

All the naturally occurring stable elements were known by the mid-twentieth century, and experiments with nuclear energy at that time brought to light a whole pantheon of heavier, short-lived radioactive elements. But only with the development of new ultra-sensitive techniques of chemical analysis have we become alerted to the complexity with which they are blended in the world, seasoning the oceans and the air with exquisite delicacy.

And so today's bottles of mineral water list their proportions of sodium, potassium, chlorine, and much else, banishing the notion that all we are drinking is H_2O. We know that elements are labile things, which is why lead water pipes and lead-based paints are no longer manufactured, and why aluminium cooking utensils are (rightly or wrongly) accused on suspicion of causing dementia. The reputations of the elements continue to be shaped by folklore and received wisdom as much as by an understanding of their quantitative effects. Is aluminium, then, good in the mineral brighteners of washing powders but bad in pots and pans? Copper salts can be toxic, but copper bracelets are rumoured to cure arthritis. We take selenium supplements to boost fertility, while selenium contamination of natural waters devastates Californian ecosystems. Which of us can say whether 0.01 milligrams of potassium in our bottled water is too little or too much?

The terminology of the elements suffuses our language, sometimes divorced from the questions of composition to which it once referred. Plumbing today is more likely to be made from plastic pipes than from the Romans' *plumbum* (lead); the lead in pencils is no such thing. 'Cadmium Red' paints often contain no cadmium at all. Tin cans have no more than the thinnest veneer of metallic tin; it is too valuable for more. The American nickel contains relatively little of that metal. And when was the last time that a Frenchman's pocketful of jingling *argent* was made of real silver?

Such are reasons why the story of the elements is not simply a tale of a hundred or so different types of atom, each with its unique properties and idiosyncrasies. It is a story about our cultural interactions with the nature and composition of matter. The Whiggish history of chemistry as a gradual elucidation and tabulation of matter's building blocks obscures a deeper and more profound enquiry into the constitution of the world, and the mutability of that constitution by human or natural agency.

Pieces of the puzzle

The concept of elements is intimately entwined with the idea of atoms, but each does not demand the other. Plato believed in the four canonical elements of antiquity, but he did not exactly concur with the notion of atoms. Other Greek philosophers trusted in atoms but did not divide all matter into a handful of basic ingredients.

Thales of Miletus (*c.*620–*c.*555 BC), one of the first known enquirers into the constitution of the physical world, posited only one fundamental substance: water. There is ample justification for this view in myth; the Hebrew god was not the only deity to bring forth the world from a primal ocean. But the Milesian school of philosophers that Thales founded produced little consensus about the *prote hyle* or 'first matter' that constituted everything. Anaximander (*c.*611–547 BC), Thales' successor, avoided the issue with his contention that things are ultimately made of *apeiron*, the 'indefinite' and unknowable first substance. Anaximenes (d. *c.*500 BC) decided that air, not water, was primary. For Heraclitus (d. 460 BC), fire was the stuff of creation.

Why should anyone believe in a *prote hyle* at all—or, for that matter, in any scheme of elements that underlies the many substances we find in the world? Why not simply conclude that rock is rock, wood is wood? Metal, flesh, bone, grass . . . there were plenty of distinct substances in the ancient world. Why not accept them at face value, rather than as manifestations of something else?

Some science historians argue that these ancient savants were searching for unity: to reduce the multifarious world to a simpler and less puzzling scheme. A predilection for 'first principles' is certainly evident in Greek philosophy, but there is also a practical reason to invoke fundamental elements: things change. Water freezes or boils away. Wood burns, transforming a heavy log to insubstantial ashes. Metals melt; food is ingested and most of it is somehow spirited away inside the stomach.

If one substance can be transformed to another substance, might that be because they are, at root, merely different forms of the same substance? The idea of elements surely arose not because philosophers were engaged on some ancient version of the physicists' quest for a unified theory but because they wanted to understand the transformations that they observed daily in the world.

To this end, Anaximander believed that change came about through the agency of contending opposite qualities: hot and cold, and dry and moist. When Empedocles (*c.*490–*c.*430 BC) postulated the four elements that gained ascendancy in Western natural philosophy, he too argued that their transformations involved conflict.

Empedocles does not exactly fit the mould of a sober and dignified Greek philosopher. Legend paints him as a magician and miracle worker who could bring the dead back to life. Reputedly he died by leaping into the volcanic maw of Mount Etna, convinced he was an immortal god. Small wonder, perhaps, that his earth, air, fire, and water were wrought into different blends—the materials of the natural world—through the agency of the colourful principles Love and Strife. Love causes mixing; Strife, separation. Their conflict is an eternal waxing and waning: at one time, Love dominates and things mix, but then Strife arises to pull them apart. This applies, said Empedocles, not just to the elements but to the lives of people and cultures.

Empedocles' four elements do not represent a multiplication of the *prote hyle*, but rather a gloss that conceals its complications. Aristotle agreed that ultimately there was only one primal substance, but it was too remote, too unknowable, to serve as the basis for a philosophy of matter. So he accepted Empedocles' elements as a kind of intermediary between this imponderable stuff and the tangible world. This instinct to reduce cosmic questions to manageable ones is one reason why Aristotle was so influential.

Aristotle shared Anaximander's view that the qualities heat, cold, wetness, and dryness are the keys to transformation, and also to our experience of the elements. It is *because* water is wet and cold that we can experience it. Each of the elements, in Aristotle's ontology, is awarded two of these qualities, so that one of them can be converted to another by inverting one of the qualities. Wet, cold water becomes dry, cold earth by turning wetness to dryness (Fig. 1).

It is tempting, and not wholly unrealistic, to regard these

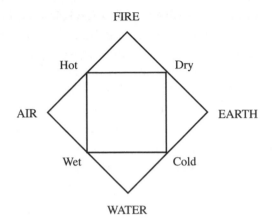

1 Aristotle believed that the four elements of Empedocles were each imbued with two qualities, by means of which they could be interconverted

ancient philosophers as belonging to a kind of gentleman's club whose members are constantly borrowing one another's ideas, heaping lavish praise or harsh criticism on their colleagues, while all the while remaining 'armchair' scientists who decline, by and large, to dirty their hands through experiment. The same image serves for those who debated the fluctuating fortunes of atoms.

Leucippus of Miletus (fifth century BC) is generally credited with introducing the concept of atoms, but we know little more about him than that. He maintained that these tiny particles are all made of the same primal substance, but have different shapes in different materials. His disciple Democritus (*c.*460–370 BC) called these particles *atomos,* meaning uncuttable or indivisible. Democritus reconciled this fledgling atomic theory with the classical elements by positing that the atoms of each element have shapes that account for their properties. Fire atoms are immiscible with others, but the atoms of the other three elements get entangled to form dense, tangible matter.

What distinguished the atomists from their opponents was not the belief in tiny particles that make up matter, but the question of what separated them. Democritus supposed that atoms move about in a void. Other philosophers ridiculed this idea of 'nothingness', maintaining that the elements must fill all of space. Anaxagoras (c.500–428 BC), who taught both Pericles and Euripides in Athens, claimed that there was no limit to the smallness of particles, so that matter was infinitely divisible. This meant that tiny grains would fill up all the nooks between larger grains, like sand between stones. Aristotle asserted—and who can blame him?—that air would fill any void between atoms. (This becomes a problem only if you consider that air is itself *made* of atoms.)

Plato had it all figured out neatly. He was not an atomist in the mould of Democritus, but he did conceive of atom-like fundamental particles of the four Empedoclean elements. His geometrical inclinations led him to propose that these particles had regular, mathematical shapes: the polyhedra called regular Platonic solids. Earth was a cube, air an octahedron, fire a tetrahedron and water an icosahedron. The flat faces of each of these shapes can be made from two kinds of triangle. These triangles are, according to Plato, the true 'fundamental particles' of nature, and they pervade all space.

The elements are converted by rearranging the triangles into new geometric forms.

There is a fifth Platonic regular solid too: the dodeca-hedron, which has pentagonal (five-sided) faces. This poly-hedron cannot be made from the triangles of the other four, which is why Plato assigned it to the heavens. There is thus a fifth classical element, which Aristotle called the aether. But it is inaccessible to earthly beings, and so plays no part in the constitution of mundane matter.

The poetic elements

The four elements of antiquity perfuse the history of Western culture. Shakespeare's Lear runs amok in the stormy rain, the rushing air, and the 'oak-cleaving thunderbolts' of fire, nature's 'fretful elements'. Two of his sonnets are paired in celebration of the quartet: 'sea and land . . . so much of earth and water wrought', and 'slight air and purging fire'. Literary tradition has continued to uphold the four ancient elements, which supply the organizing principle of T. S. Eliot's *Quartets*.

The Greek philosophers coupled a four-element theory to the idea of four 'primary' colours: to Empedocles these were white, black, red, and the vaguely defined *ochron*, consistent with the preference of the classical Greek painters for a four-colour palette of white, black, red, and yellow. The Athenian astrologer Antiochos in the second century AD assigned these colours, respectively, to water, earth, air, and fire.

A determination to link the four elements to colours

persisted long after the Greek primaries had been discarded. The Renaissance artist Leon Battista Alberti awarded red to fire, blue to air, green to water, and 'ash colour' (*cinereum*) to earth; Leonardo da Vinci made earth yellow instead. These associations would have surely informed the contemporaneous ideas of painters about how to mix and use colours.

This fourness of fundamental principles reaches further, embracing the four points of the compass (Chinese tradition acknowledges five elements, and five 'directions') and the four 'humours' of classical medicine. According to the Greek physician Galen (AD *c*.130–201), our health depends on the balance of these four essences: red blood, white phlegm, and black and yellow bile.

Even allowing for the ancient and medieval obsession with 'correspondences' among the characteristics and creations of nature, there is clearly something about the four Aristotelian elements that has deep roots in human experience. The Canadian writer Northrop Frye writes: 'The four elements are not a conception of much use to modern chemistry—that is, they are not the elements of nature. But ... earth, air, water and fire are still the four elements of imaginative experience, and always will be.'

This is why the French philosopher Gaston Bachelard felt it appropriate to explore the 'psychoanalytic' influence of these elements (in particular water and fire) in myth and poetry.

I believe it is possible [he said] to establish in the realm of the imagination, a *law of the four elements* which classifies various kinds of material imagination by their connections with fire, air, water or earth ... A material element must provide its own substance, its particular rules and poetics. It is not simply coincidental that primitive philosophies often made a decisive choice along these lines. They associated with their formal principles one of the four fundamental elements, which thus became signs of *philosophic disposition.*

Bachelard suggests that this disposition is, for every individual, conditioned by his or her material environment:

the region we call home is less expanse than matter; it is granite or soil, wind or dryness, water or light. It is in it that we materialize our reveries, through it that our dream seizes upon its true substance. From it we solicit our fundamental colour. Dreaming by the river, I dedicated my imagination to water, to clear, green water, the water that makes the meadows green.

Despite a tendency to overestimate the primacy of the four-element scheme—there have been, as we have seen, many others—this idea goes some way towards explaining the longevity of Empedocles' elements. They *fit*, they accord with our experience. They distinguish different *kinds* of matter.

What this really means is that the classical elements are familiar representatives of the different *physical states* that matter can adopt. Earth represents not just soil or rock, but all solids. Water is the archetype of all liquids; air, of all gases and vapours. Fire is a strange one, for it is indeed a unique and striking phenomenon. Fire is actually a dancing plasma of molecules and molecular fragments, excited into a glowing state by heat. It is not a substance as such, but a variable combination of substances in a particular and unusual state caused by a chemical reaction. In experiential terms, fire is a perfect symbol of that other, intangible aspect of reality: light.

The ancients saw things this way too: that elements were *types*, not to be too closely identified with particular substances. When Plato speaks of water the element, he does not mean the same thing as the water that flows in rivers. River water is a manifestation of elementary water, but so is molten lead. Elementary water is 'that which flows'. Likewise, elementary earth is not just the stuff in the ground, but flesh, wood, metal.

Plato's elements can be interconverted because of the geometric commonalities of their 'atoms'. For Anaxagoras, all material substances are mixtures of all four elements, so one substance changes to another by virtue of the growth in proportion of one or more elements and the corresponding diminution of the others. This view of matter as intimate blends of elements is central to the antiquated elementary theories, and is one of the stark contrasts with the modern notion of an element as a fundamental substance that can be isolated and purified.

Age of metals

With Aristotle's endorsement, the Empedoclean elements thrived until the seventeenth century. With that blessing withheld, atomism withered. The Greek philosopher Epicurus (341–270 BC) established an atomistic tradition that was celebrated in 56 BC by the Roman poet Lucretius in his tract *De rerum natura* (*On the Nature of Things*). This atomistic poem was condemned by religious zealots in the Middle Ages, and barely escaped complete destruction. But it surfaced in the seventeenth century as a major influence on the French scientist Pierre Gassendi (1592–1655), whose vision of a mechanical world of atoms in motion represented one of the many emerging challenges to the Aristotelian orthodoxy.

Not everyone was ready for such radical changes. Gassendi's fellow 'mechanist' Marin Mersenne (1588–1648), in many ways a progressive thinker, nevertheless endorsed the Condemnation of 1624 in which Étienne de Clave was arrested, claiming that such gatherings encouraged the propagation of 'alchemical' ideas. Alchemy, however, had plenty more to say about the elements.

It may seem strange from today's perspective that several of the substances recognized today as elements—the metals gold, silver, iron, copper, lead, tin, and mercury—were not classed as such in antiquity, even though they could be prepared in an impressively pure state. Metallurgy is one of the most ancient of technical arts, and yet it impinged relatively little on the theories of the elements until after the Renaissance. Metals, with the exception of fluid mercury, were considered simply forms of Aristotelian 'earth'.

Alchemy, which provided the theoretical basis for metallurgy, gradually changed this. It added a deeper sophistication to ideas about the nature and transformation of matter, providing a bridge between the old and new conceptions of the elements.

If the notion of a single *prote hyle* was initially something of a dead end for a theory of matter, the Aristotelian elements were not a great deal better. The differences between lead and gold mattered very much to society, but the four-element theory could say little about them. A more refined scheme was needed to account for the metals.

Gold and copper are the oldest known metals, since they occur in their pure, elemental forms in nature. There is evidence of the mining and use of gold in the region of Armenia and Anatolia from before 5000 BC; copper use is similarly ancient in Asia. Copper mostly occurs not as the metal, however, but as a mineral ore: a chemical compound of copper and other elements, such as copper carbonate (the minerals malachite and azurite). These copper ores were used as pigments and colouring agents for glazes, and it is likely that copper smelting, which dates from around 4300 BC, arose from a happy accident during the glazing of stone ornaments called faience in the Middle East. The synthesis of bronze, an alloy of copper and tin, dates from about the same time.

Lead was smelted from one of its ores (galena) since around 3500 BC, but was not common until 1,000 years later. Tin seems to originate in Persia around 1800–1600 BC, and iron in Anatolia around 1400 BC. This sequence of discovery of the metals reflects the degree of difficulty in separating the pure metal from its ore: iron clings tightly to oxygen in the common mineral ore haematite (ochre), and intense heat and charcoal are needed to prise them apart.

With this profusion of metals, some scheme was needed to classify them. Convention dictated that this be at first a system of correspondences, so that the seven known metals became linked with the seven known celestial bodies and the seven days of the week (Table 1). Since all metals shared attributes in common (shininess, denseness, malleability), it seemed natural to suppose that they were different only in degree and not in kind. Thus arose the precept that metals 'mature' in the earth, beginning with dull, dirty lead and culminating in glorious gold.

TABLE 1 *The seven 'classical' metals and their correspondences*

Metal	Celestial body	Day
Gold	Sun	Sunday
Silver	Moon	Monday
Mercury (Quicksilver)	Mercury	Wednesday (Fr. *Mercredi*)
Copper	Venus	Friday (Fr. *Vendredi*)
Iron	Mars	Tuesday (Fr. *Mardi*)
Tin	Jupiter	Thursday (Fr. *Jeudi*)
Lead	Saturn	Saturday

This was the central belief of alchemy. If metals may indeed be interconverted one to another in the deep earth, perhaps the alchemist could find a way to accelerate the process artificially and make gold from baser metals. But how was this done?

Attempts to transmute other metals to gold may have been made as long ago as the Bronze Age. But after the eighth century AD they were no longer haphazard; they had a theoretical underpinning in the sulphur-mercury theory of the Arabic alchemist Jabir ibn Hayyan. Jabir is more the name of a school of thought than of a person. Many more writings are attributed to him than he could possibly have written, and there is some doubt about whether he existed at all. The Jabirian tradition works curious things with the Aristotelian elements. It accepts them implicitly but then, so far as metals are concerned, adds another layer between these fundamental substances and reality.

According to Jabir, the 'fundamental qualities' of metals are the Aristotelian hot, cold, dry, and moist. But the 'immediate qualities' are two 'principles': sulphur and mercury. All metals are deemed to be mixtures of sulphur and mercury. In base metals they are impure; in silver and gold they attain a higher state of purity. The purest mixtures of this sulphur and mercury yield not gold but the Holy Grail of alchemy, the Philosopher's Stone, the smallest quantity of which can transform base metals to gold.

Some scholars have identified Jabir's sulphur and mercury with the Aristotelian opposites fire and water. One thing is sure: they are not the yellow sulphur and the glistening, fluid mercury of the chemistry laboratory, which were known in more or less pure form even to the alchemists. Instead, these two principles were rather like the four classical elements: 'ideal' substances embodied only imperfectly in earthly materials.

So the Jabirian system embraced the four classical elements and then buried them, just as the Aristotelian elements allowed but ignored the universal *prote hyle*. It marks the beginning of a tendency to pay lip service to Aristotle while getting on with more practical concerns about what things are made of.

The next step away from the traditions of antiquity involved the addition of a third 'principle' to Jabir's sulphur and mercury: salt. Whereas the first two were components of metals, salt was considered an essential ingredient of living bodies. In this way alchemical theory became more than a theory of metallurgy and embraced all the material world. The three-principle theory is generally attributed to the Swiss alchemist Paracelsus (1493–1541), although it is probably older. Paracelsus asserted that sulphur, salt, and mercury 'form everything that lies in the four elements'.

So these Paracelsian principles were not meant to be elements in themselves, but rather a material manifestation of the ancient elements. By the end of the seventeenth century, things had moved on again. There was no longer any perceived obligation to square one's views with Aristotle, and the 'principles' were widely regarded as elements in their own right. Jean Béguin listed a popular scheme of five elements: mercury, sulphur, salt, phlegm, and earth. He claimed that none of them was pure—each contained a little of the others.

Johann Becher (1635–c.1682), an influential German alchemist of the most flamboyant kind, accepted that air, water, and earth were elements, but did not accord them equal status. Air, he believed, was inert and did not take part in processes of transformation. He felt that the differences between the many dense substances of the world stemmed from three different types of earth. *Terra fluida* was a fluid element that gave metals their shininess and heaviness. *Terra pinguis* was a 'fatty earth', abundant in organic (animal and vegetable) matter, which made things combustible. *Terra lapidea* was 'vitreous earth', which made things solid. These three earths are in fact nothing but mercury, sulphur, and salt in disguise, but we will see later how modern chemistry arose out of them.

The sceptical chymist

The impetus for this sudden profusion and elaboration of elemental schemes came mostly from experiment. No longer content to apportion matter into the abstract, remote elements of the Greeks, the early chemists of the seventeenth century began trying to understand matter by practical means.

Alchemy always had a strong experimental side. In their endless quest for the Philosopher's Stone, alchemists burnt, distilled, melted, and condensed all manner of substances and stumbled across many technologically important new compounds, such as phosphorus and nitric acid. But in the 1600s there appeared a transitional group of natural philosophers whose primary objective was no longer to conduct the Great Work of alchemical transformation but to study and understand matter at a more mundane level. These 'chymists' were neither alchemists nor chemists; or, rather, they were a bit of both. One of them was Robert Boyle (1627–91).

The Eton-educated son of an Irish aristocrat, Boyle became part of the innermost circle of British science in the mid-seventeenth century. He was on good if not intimate terms with Isaac Newton (hardly anyone was intimate with Newton), and was involved in the founding of the Royal Society in 1661. Like many of his contemporaries, he was passionately interested in alchemy; but, crucially, he was also an independent and penetrating thinker.

Traditionally portrayed as a broadside against alchemy in general, Boyle's classic book *The Sceptical Chymist* (1661) in fact aims to distinguish the learned and respectable alchemical 'adepts' (such as Boyle himself) from the 'vulgar laborants' who sought after gold by means of blind recipe following. The book's lasting value to chemistry comes from Boyle's assault on all the main schools of thought about the elements. These, he said, are simply incompatible with the experimental facts.

The conventional four-element theory claimed that all four of Aristotle's elements are present in all substances. But Boyle observes that some materials cannot be reduced to the classical elementary components, however they are manipulated by 'Vulcan', the heat of a furnace:

> Out of some bodies, four elements cannot be extracted, as Gold, out of which not so much as any *one* of them hath been hitherto. The like may be said of Silver, calcined Talke [roasted talc], and divers other fixed bodies, which to reduce into four heterogeneal substances, is a taske that has hitherto proved too hard for Vulcan.

In other words, elements are to be found not by theorizing but by experiment: 'I must proceed to tell you that though the assertors of the four elements value reason so highly . . . no man had ever yet made any sensible trial to discover their number.'

Boyle's definition of an element is nothing very controversial by the standards of the times:

> certain primitive and simple, or perfectly unmingled bodies; which not being made of any other bodies, or of one another, are the ingredients of which all those called perfectly mixt bodies are immediately compounded, and into which they are ultimately resolved.

But he then proceeds to question whether anything of this sort truly exists—that is, whether there are elements at all. Certainly, Boyle holds back from offering any replacement for the elemental schemes he demolishes, although he shows some sympathy for the idea, advocated by the Flemish scientist Johann Baptista van Helmont, that everything is made of water.

By the end of the seventeenth century, then, scientists were not really any closer to enumerating the elements than were the Greek philosophers. Yet a hundred years later the British chemist John Dalton (1766–1844) wrote a textbook that outlined a recognizably modern atomic theory and gave a list of elements that, while still very incomplete and sometimes plain wrong, is in content and in spirit a clear precursor to today's tabulation of the hundred and more elements. Why had our understanding of the elements changed so fast?

Boyle's demand for experimental analysis as the arbiter of elemental status is a central component of this change. Another reason for the revolution was the relinquishment of old preconceptions about what elements should be like. For the classical scholars, an element had to correspond to (or at least be recognizable in) stuff that you found around you. Many of the substances today designated as elements are ones almost all of us will never see or hold; in antiquity, that would seem an absurd complication. (True, no one could hold the aether, but everyone could see that the heavens sat over the earth.) Some confusion was also dispelled as scientists began to appreciate that substances could change their physical state—from solid to liquid to gas—without changing their elemental composition. Ice is not water turned to 'earth'—it is frozen water.

In short, there is nothing *obvious* about the elements. Until the twentieth century, scientists had no idea why there should be so many, nor indeed why there should not be thousands more. The elements cannot be deduced by casual inspection of the world, but only by the most exacting scrutiny using all the complicated tools of modern science.

This is why, perhaps, some people would like to stick with earth, air, fire, and water. They are not the elements of chemistry, but they say something resonant about how we interact with the world and about the effect that matter has on us.

Revolution

How Oxygen Changed the World

It is often said that Antoine Laurent Lavoisier did for chemistry what Isaac Newton did for physics and Charles Darwin for biology. He transformed it from a collection of disparate facts into a science with unified principles.

But timing is crucial. Newton's work in the seventeenth century signals the beginning of the Enlightenment, the confidence in rationalism as a way both to understand the universe and to improve the human condition. Darwin's theories began to take hold as the solid certainties of nineteenth-century science and culture gave way before the giddy perspectives of modernism; all the old rules of art, music, and literature were changing at the same time.

And Lavoisier? His was the fate of the Enlightenment's brave new world: slaughtered during Robespierre's Reign of Terror. The liberal optimism of philosophers and thinkers like Voltaire, Montesquieu, and Condorcet foundered before the fickle passions and arbitrary brutality of the French Revolutionaries. Reason was overthrown, and, in the decades that followed, chemistry became the supremely Romantic science.

Lavoisier (1743–94), like Condorcet, was misfortunate that the leading thinkers in France were likely, sooner or later, to become embroiled in politics. Whereas in England science was still the pursuit of 'gentlemen' with money and leisure to spare, France had its state-approved Academy of Sciences whose members commonly filled public offices and became highly visible figures in political life (Fig. 2).

Lavoisier was a tax collector before he became a famous scientist, and that was largely what sealed his fate. But his chemical expertise also secured him the prominent position of director on Louis XVI's Gunpowder Administration, and as treasurer and effective secretary of the Academy of Sciences he vigorously opposed its dissolution by the anti-elitist Jacobin administration in 1793. Lavoisier was a sitting target for the Revolutionary witch-hunters, who were determined to purge the nation of anyone whose loyalty to the Republic they found reason to doubt. That is why, in 1794, Lavoisier was forced to bow his head to the blade that had just removed his father-in-law's.

Two centuries later, the debate still rages about whether Lavoisier was or was not the true discoverer of one of chemistry's most important elements: oxygen. It has become the subject of a play written by two of the world's leading chemists, the Nobel laureate Roald Hoffmann and the co-inventor of the contraceptive pill, Carl Djerassi. In *Oxygen*, the Nobel Committee of 2001 has decided to award 'retro-Nobel' prizes for great discoveries made before the prize was inaugurated

2 Antoine Laurent Lavoisier (1743–94), the 'Newton of chemistry', and his wife and sometime assistant Marie Anne Lavoisier

in 1901. They decide that the first chemistry prize must go to oxygen's discoverer, because, says one of the characters, 'the Chemical Revolution came from oxygen'. Lavoisier gave the element its name, but he was certainly not the first to make it, nor to recognize it as a distinct and important substance. The Nobel Committee argues furiously over the leading three candidates, while a fictional encounter between them in 1777 reveals new insights into their own struggles to secure priority.

Yet that is only part of the tale. Oxygen provides not only the central organizing principle for modern chemistry but a bridge between the new and the old, between the alchemical roots of Robert Boyle's 'chymistry' and the syntheses of endless wonders in today's chemical plants. In joining the two, it marks a crucial stage in the developing concept of an element.

Something in the air

Lavoisier delivered two shocks to the Aristotelian elements. His experiments on water led him to conclude in 1783 that it 'is not a simple substance at all, not properly called an element, as had always been thought'. And, concerning that other fluid element of antiquity, he announced that 'atmospheric air is composed of two elastic fluids of different and opposite qualities', which he called 'mephitic air' and 'highly respirable air'. Neither water nor air, in other words, is an element.

He named the constituents of water hydrogen ('water-former') and oxygen, which combine in a two-to-one ratio reflected in the familiar chemical formula H_2O. Air is a more complex substance. The fraction that is 'highly respirable air', Lavoisier realized, is an element in itself: oxygen. The name comes from the Greek for 'acid-former', as Lavoisier wrongly believed that oxygen was a component of all acids. For the 'fluid' that Lavoisier called mephitic air he proposed the name *azot* or *azotic gas*, a Greek term indicating that it is inimical to life. Lavoisier found that, when he isolated this component, it had the 'quality of killing such animals as are forced to breathe it'. Reasonably enough, he concluded that it was noxious. In fact it is not poisonous but simply useless: separated from oxygen, it cannot sustain life. Lavoisier noted that this gas 'is proved to form a part of the nitric acid, which gives a good reason to have called it *nitrigen*'. He preferred his *azot*, however, and so did the other French chemists—which is why nitrogen is known to this day as *azote* in France.

Lavoisier was not intent on wholly demolishing tradition, however, vouching that: 'We have not pretended to make any alteration upon such terms as are sanctified by ancient custom; and therefore . . . retain the word *air*, to express that collection of elastic fluids which composes our atmosphere.'

His assessment of this 'collection of fluids' was somewhat incomplete, although understandably so. Oxygen and nitrogen between them account for 99 per cent of air; but the remainder is a fantastic blend. Mostly it is argon (see page 193), an extremely unreactive element. There is a small, variable proportion of water vapour (enough to condense into clouds and raindrops when air is cooled), and about 0.08 per cent of air is carbon dioxide. Other trace gases include methane, nitrous oxide, carbon monoxide, sulphur dioxide, and ozone. Until the past few decades, many of the minor constituents of air went undetected. But, despite their low concentrations, they play a crucial role in atmospheric and environmental chemistry. Some are greenhouse gases, warming the planet. Others are toxic pollutants. Some have natural sources; others are solely human made; many are both. To understand the properties and behaviour of the atmosphere, chemists commonly now have to take into account reactions involving dozens or even hundreds of trace gases and their offspring.

Oxygen and nitrogen are elements, but most of these other gases are *compounds* formed by the reaction and joining together of two or more different elements. In oxygen gas, each atom of oxygen is bound to another atom of oxygen. In carbon monoxide, an oxygen atom is linked to an atom of carbon.

Somewhat confusingly, when chemists use the term 'element', they can thus be referring either to a specific kind of atom—oxygen in rust or water is still an element in this sense—or to a physical substance containing only one kind of atom, like oxygen gas or a piece of ruddy copper metal. Some elements, including most metals, are usually found naturally in compounds, in which their atoms are linked to those of other elements. Other elements occur naturally in a pure or 'elemental' form, like sulphur or gold. It is not dissimilar to saying that a cat is both an abstract thing with distinguishing properties—pointed ears, a tail, a tendency to purr and chase mice—and the very real, warm ginger creature that sits at our hearth.

So air is (mostly) oxygen and nitrogen; water is oxygen and hydrogen. But the elements that constitute air do not form the same kind of mixture as those in water. Chemical bonds link each atom of oxygen to two atoms of hydrogen in water, and only a chemical reaction will separate them. In air, the two elements are just mixed physically, like grains of sand and salt. They can be separated without a chemical reaction. In practice, Lavoisier found it necessary to use a chemical reaction to perform the separation: he allowed the oxygen to combine with other substances through combustion, leaving behind almost pure nitrogen. But modern techniques can perform the physical separation of these elements.

Oxygen's shadow

Lavoisier's conclusion about air was not new. Just as he was not the first to make water from its component elements, neither could he lay claim to the priority for deducing that air contains two dissimilar substances. What was special about Lavoisier's claim was not the observation but the interpretation.

The second half of the eighteenth century was the age of 'pneumatick chemistry', when the properties of gases, typically called 'airs', were the focus of the discipline. The invention of the pneumatic trough, a device for collecting gases emanating from heated substances, by the English clergyman Stephen Hales in the early part of the century, was pivotal for bringing about this emphasis. Whereas in antiquity 'air' implied anything gaseous, Hales's apparatus allowed chemists to appreciate that not all such 'emanations' were alike, and so could not justifiably be regarded as the same unadulterated element.

There was, for example, the 'fixed air' studied by Scottish chemist Joseph Black (1728–99). In the 1750s, Black found that a gas was produced when carbonate salts were heated or treated with acid. The air, he reasoned, was 'fixed' in the solids until liberated. Unlike common air, fixed air turned lime water (a solution of calcium hydroxide) cloudy. We now recognize that this is due to the formation of insoluble calcium carbonate—basically chalk. Black found that human breath, the gases given off during combustion, and the gaseous product of fermentation, all have the same effect on lime water. This fixed air is carbon dioxide, into which carbonates decompose when heated.

Black's student Daniel Rutherford (1749–1819) called this gas 'mephitic air' instead: *mephitis* is a noxious emission in legend, thought to emanate from the earth and cause pestilence. It seemed an apt name, for animals died in an atmosphere of this new gas. Rutherford's 'air' is not, however, the same as Lavoisier's mephitic air, which is nitrogen. Yet Rutherford is himself credited with discovering nitrogen, for he found that it is an unreactive component of common air. Only about a fifth of common air is 'good', supporting life, Rutherford reported in 1772. If this good air is consumed in some way, that which remains extinguishes candles and suffocates mice. Two other English pneumatick chemists, Henry Cavendish (1731–1810) and Joseph Priestley (1733–1804), made the same observations in the 1760s; indeed, similar results date back to the time of Robert Boyle. But Black was the first (marginally) to advance the notion that nitrogen, as it later became known, was a separate element.

Joseph Priestley's experiments with Hales's trough were phenomenally fertile. He isolated around twenty different airs, including hydrogen chloride, nitric oxide, and ammonia. But neither he nor any of his contemporaries regarded these substances initially as distinct compounds in their own right. The legacy of Aristotle's elements was still strong, and the pneumatick chemists preferred to regard each gas as 'common air' altered in some manner—for example, in states of greater or lesser impurity. Even Lavoisier found this a hard habit to shake off.

This prejudice reflected more than an allegiance to classical ideas, however. The pneumatick chemists had a theory to explain the chemical reactions of gases, which they moulded to fit every new observation. It invoked chemistry's most notorious pseudo-element: phlogiston.

Alchemy mutated into modern chemistry in several stages, and the phlogiston theory was arguably the last of them. We can trace this hypothetical substance back to Jabir ibn Hayyan's sulphur, a supposed component of all metals. *Real* sulphur, the yellow solid mined from the earth, was a combustible substance, a component of gunpowder and the brimstone that bubbles beneath the fires of hell. So it is understandable how the alchemical sulphur of the three Paracelsian 'principles' became Johann Becher's *terra pinguis*: fatty earth, the oily principle of combustibility. Becher's disciple Georg Ernst Stahl (1660–1734) gave it a new name: phlogiston, from the Greek word for 'to burn'.

To some chemists phlogiston was fire itself: a form of the ancient element. Others, accepting the blurring of the demarcation between 'elements' and alchemical 'principles', concurred with Becher's definition of *terra pinguis*: 'Metals contain an inflammable principle which by the action of fire goes off into the air.'

It seems reasonable enough to assume from the flames and smoke dancing above a burning log that the wood is releasing some substance into the air. This, then, was phlogiston, the essence of flammability. You want proof? Burn a candle flame in a sealed container. The flame dies out eventually, said phlogistonists, because the air has become saturated with phlogiston given off by the candle and can receive no more.

Metals do not generally burn with a bright flame, but when they are heated in air they can be converted into new, dull substances. This process was called calcination in the eighteenth century, and the products were calxes. If a calx is heated in the presence of charcoal, the metal is recovered. It was assumed that metals too give out phlogiston during calcination. Charcoal was deemed to be rich in phlogiston (why else would it burn so well in ovens and furnaces?), and so it was capable of restoring this substance to the calx, regenerating the metal.

There is just one problem. It is true that wood, losing mass as it burns, seems to be giving out some substance into the air. But calcined metals *gain* weight. How can they get heavier by losing phlogiston? Most chemists ducked the issue; some asserted that phlogiston was weightless, or even that it had negative weight or the ability to convey buoyancy.

Stahl's phlogiston theory was elaborated to explain not just combustion but many other processes, including biological ones. It accounted for acids and alkalis, for respiration and the smells of plants. It was a chemical theory that, if not all embracing, at least gave the discipline an impressive unity.

In 1772 Lavoisier was still a believer in the phlogistonic orthodoxy. But he had begun to doubt that this was all there was to combustion. He proposed towards the end of that year that metals take up ('fix') air when calcined, and that the calx releases this fixed air when 'reduced' back to metals with the agency of charcoal and heat. Hearing of Black's fixed air in 1773, he decided that this was what metals combine with to form a calx. That at least explained the gain in weight. It also weakened the need to invoke phlogiston at all.

Then a French pharmacist named Pierre Bayen pointed out to Lavoisier that 'calx of mercury', which we would now call mercuric oxide, can be converted to mercury simply by heating, without the need for 'phlogiston-rich' charcoal. Moreover, the gas released in this process was not Black's fixed air, but something quite different. What was this gas?

That started to become clear to Lavoisier when Joseph Priestley came to dinner.

Priestley, a nonconformist Presbyterian minister, was supported in his scientific studies by the patronage of the Earl of Shelburne, in whose house Priestley was tutor. In August 1774 Priestley conducted the same experiment as Bayen, heating mercuric oxide and collecting the gaseous product. He found that a candle flame placed in this gas burned even more brightly than in common air, and that a lump of smouldering charcoal became incandescent. In this 'air', combustion thrived.

Obviously, thought Priestley, the 'air' was peculiarly lacking in phlogiston, and was therefore especially avid to soak it up from burning substances. Priestley never swayed from his firm conviction in the phlogiston theory as long as he lived, and he called his new gas 'dephlogisticated air'.

In 1775 Priestley discovered it had an even more miraculous property. Mice placed in a glass vessel full of 'dephlogisticated air' survived for much longer than mice in an identical vessel containing common air. There was something *vital* about this substance, and, when Priestley himself inhaled it, he reported that 'my breath felt peculiarly light and easy for some time afterward'. He envisaged that it might become used as a health-enhancing substance, although 'hitherto only my mice and myself have had the privilege of breathing it'.

Here Priestley may have been mistaken, for in 1674 Robert Boyle's assistant John Mayow (1641–79) asserted that a gas released by heating nitre (potassium nitrate) turned arterial blood red in the lungs. Mayow asserted that metals gained weight during calcination because of the uptake of particles of this 'nitro-aerial' gas (which was, of course, nothing other than oxygen). And in 1771-2 a Swedish apothecary named Carl Wilhelm Scheele, one of the finest experimental chemists of his age, performed the same experiment as Mayow and isolated a gas that enhanced burning. He supposed that this flammable component of common air, which he called 'fire air', combined with phlogiston during burning.

So Priestley's dephlogisticated air had a hidden past. Scheele's work was still unknown in 1775, since the apothecary did not announce his findings (which included the fact that 'fire air' comprised one-fifth of common air) until 1777.

Priestley and Shelburne dined with Lavoisier in Paris in October 1774, and Priestley mentioned his findings at the table. Together with Bayen's results, this persuaded Lavoisier that metals were not after all combining with 'fixed air' to form calxes. Bayen reported only that the gas released from mercuric oxide was like common air; and in March 1775 Lavoisier announced that his own experiments with mercuric oxide revealed all calxes to be a combination of metals with such a gas.

Seeing this report, Priestley realized that Lavoisier had not quite appreciated the 'superior' qualities of his 'dephlogisticated air'—it was not merely common air. He sent the Frenchman a sample of the gas to verify that this was so. As a result, Lavoisier presented a paper to the French Academy in April in which he identified the principle of combustion—Priestley's gas—as an especially 'pure air'. In keeping with his notorious arrogance, he made no mention of the contributions of Priestley and Bayen.

Priestley, Lavoisier, and Scheele feature in the play *Oxygen* as the three contenders for the discovery of oxygen. Scheele's part in the real drama was not quite so isolated as it might appear. His account of the discovery was sent to the publishers in 1775, but took two years to appear in print. More significantly, Scheele sent a letter to Lavoisier in September 1774 outlining his findings. The fate of the letter is not known; but in *Oxygen* it becomes a central part of the plot.

Lavoisier may have been cavalier with his treatment of priority issues, but he went far beyond replicating the results of others. To Priestley, oxygen was always going to be regarded as a form of common air modified by the removal of phlogiston; Scheele too saw things very much in these terms. Lavoisier came to understand that this 'pure air' was actually a substance in its own right. In that case, air itself was not elemental but a mixture. It is Lavoisier who made oxygen an element.

The chronology of events suggests that oxygen arose purely from attempts to explain combustion. But Lavoisier was equally keen to make this new element the explanatory principle of acidity, itself still a profound mystery to chemists. In this he was less successful. Many non-metallic elements, such as sulphur, carbon, and phosphorus, combine with oxygen to produce gases that dissolve in water to make acids, and that is why Lavoisier named the new element as he did (in German oxygen is still known as *Sauerstoff*, 'acid stuff'). But not all acids contain oxygen; and those that do, do not derive their acidity from it.

Lavoisier's belief reveals that he still held a somewhat traditional view of elements. They were generally regarded as being rather like colours or spices, having intrinsic properties that remain evident in a mixture. But this is not so. A single element can exhibit very different characteristics depending on what it is combined with. Chlorine is a corrosive, poisonous gas; combined with sodium in table salt, it is completely harmless. Carbon, oxygen, and nitrogen are the stuff of life, but carbon monoxide and cyanide (a combination of carbon and nitrogen) are deadly. This was a hard notion for chemists to accept. Lavoisier himself came under attack for claiming that water was composed of oxygen and hydrogen: for water puts out fires (it is 'the most powerful antiphlogistic we possess', according to one critic), whereas hydrogen is hideously flammable.

The discovery of oxygen did not just make phlogiston redundant; the two were fundamentally incompatible. Oxygen is the very opposite of phlogiston. It is consumed during burning, not expelled. Burning ends when the air is devoid of oxygen, not when it is saturated with phlogiston. Indeed, it is this mirror-image quality that made phlogiston seem to work so well. Science needed an element like this to explain combustion—but it simply looked at the problem the wrong way round. Phlogiston was oxygen's shadow.

But Lavoisier rejected it only in stages. At first he simply avoided mentioning it. Not until 1785 was he prepared to issue a formal denunciation. When it came, however, it was harsh:

> Chemists have made phlogiston a vague principle, which is not strictly defined and which consequently fits all the explanations demanded of it. Sometimes it has weight, sometimes it has not; sometimes it is free fire, sometimes it is fire combined with an earth; sometimes it passes through the pores of vessels, sometimes they are impenetrable to it. It explains at once causticity and non-causticity, transparency and opacity, colour and the absence of colours. It is a veritable Proteus that changes its form every instant!

Yet even Lavoisier could not quite relinquish everything that phlogiston stood for. Like many of his contemporaries, he regarded heat as a physical substance, rather like the ancient elemental fire. He called it caloric, and it sounded suspiciously like phlogiston in another guise. Caloric was what made substances gaseous; oxygen gas was replete with it. When oxygen reacted with metals to form calxes, caloric was released (heat was given out), and in consequence the oxygen became dense and heavy.

These ideas are evident in an essay of Lavoisier's from 1773, in which he identifies the three different physical states of matter: solid, liquid, and gas. Here he makes the crucial distinction between the physical and chemical nature of substances, which confused the ancients and led to their minimal elemental schemes. 'The same body', says Lavoisier, 'can pass successively through each of these states, and in order to make this phenomenon occur it is necessary only to combine it with a greater or lesser quantity of the matter of fire.'

Belief in a kind of elemental fire (even with a fancy new name) is not the only remnant of the classical past in Lavoisier's view of the elements. He retained the notion that true elements are ubiquitous, or are at least components of very many substances:

> it is not enough for a substance to be simple, indivisible, or at least undecomposed for us to call it an element. It is also necessary for it to be abundantly distributed in nature and to enter as an essential and constituent principle in the composition of a great number of bodies.

Despite these throwbacks, Lavoisier transformed the way chemists thought about elements. At the beginning of the eighteenth century it was common to imagine just five of them. In 1789 Lavoisier consolidated his oxygen theory by publishing a textbook, *Traité élémentaire de chimie* (*An Elementary Treatise on Chemistry*), that defined an element as any substance that could not be split into simpler components by chemical reactions. And he listed no fewer than thirty-three of them. It would require nineteenth-century physics to show that some of these were fictitious (light, caloric). Several others were in fact compounds that chemists had not yet found how to decompose to their elements. But the message was clear: there is no 'simple' scheme of elements. There are lots of them out there, and it was up to chemists to find them.

Signs of life

Scientists have recently gained their first glimpse of a planet outside our solar system. The first 'extrasolar' planet was detected in 1996 by the wobble that it transmits to its mother star as it circulates in orbit. But in 1999 astronomers were able to detect the light reflected from such a planet. It was slightly blue.

Sadly, this does not mean the planet is like Earth; the blue tint probably comes from other gases in the planet's atmosphere. But what if one day scientists find a planet whose reflected light contains the telltale fingerprint of oxygen, as does that of our own world? Then it will be hard to conclude other than that the planet harbours life.

This seems a big leap: why does oxygen imply life? Until the 1960s, scientists tended to believe that the Earth's oxygen-rich atmosphere—it is roughly one-fifth oxygen and four-fifths nitrogen—was a 'given', a result of geological processes on the early Earth. According to this picture, a planet with an oxygen blanket could support life but does not necessarily do so.

Now they see things very differently. The chemical composition of the air is not a precondition for life but the result of it. Around two billion years ago, primitive living organisms transformed the atmosphere from one largely devoid of oxygen to one with plenty of it.

There is no known geological process that can maintain a high level of oxygen in our planet's atmosphere. Eventually the gas will react with rocks and become locked away in the ground. Only biological processes can strip oxygen out of its combinations with other elements and return it to the skies. If all life on Earth were to end, the oxygen level would gradually dwindle to insignificance. For this reason, an oxygen-rich atmosphere is a beacon that proclaims the presence of life beneath it.

All animals rely on oxygen, but there is nothing very surprising about organisms that do not. Many bacteria are anaerobic: they do not consume oxygen, and indeed are averse to it. These organisms thrive in the mud of the seabed and of marshlands, in deep oilfields, and many other places where air does not penetrate.

When life began, over 3.8 billion years ago, the first cells were anaerobic. The atmosphere at that time was probably a mixture of nitrogen with gases such as carbon monoxide and water vapour, or perhaps methane. Like any other organisms, these primitive bacteria needed some source of energy to drive their biochemical processes, and some researchers believe they may have at first found this source in the heat and chemical energy of undersea volcanoes.

But there is a more widespread and abundant energy source: sunlight. At some stage in early evolution, life discovered how to harness the sun's rays through photosynthesis. The light energy is used to split apart carbon dioxide and synthesize the carbon-based molecules of life. A by-product of the photosynthetic reactions of most organisms is oxygen. For millions of years this gas was soaked up by other substances, such as iron that was dissolved in the seas. But eventually these 'oxygen sinks' were all used up and and oxygen began to accumulate in the atmosphere.

To us this sounds fortuitous, but to photosynthetic cells it was the biggest outbreak of global pollution the world has ever seen. To them oxygen was sheer poison. It is perceived as a friendly element, but it is actually one of the most corrosive and destructive. Oxygen's urge to engage in chemical reactions is excelled by only a very few other elements.

After all, it takes only a single spark to persuade an entire forest to react with oxygen. The consequence in 1998–9 was a haze of smoke that covered Indonesia and altered the local climate. There is geological evidence for global wildfires in the distant past that make this one seem like a bonfire.

St Matthew warns that there are no treasures on Earth but that 'moth and rust doth corrupt'—for there was until recently no way to protect gleaming iron and steel from the avid combining power of oxygen. It turns old paintings brown as it transforms the varnish; exposed to air, most metals develop a rind of oxide within seconds.

Nature, however, makes do. If the air is full of poison, it will learn to live on poison. We breathe oxygen not because it is inherently good for us but because we have evolved ways of making it less bad for us. Enzymes mop up the deadly compounds formed as oxygen is used to burn sugar in the energy factories of our cells. These compounds include hydrogen peroxide, used as an industrial and a domestic bleach, and the even more destructive superoxide free radical. Such substances damage the delicate biomolecules of our cells, including DNA. Cells have molecular mechanisms that strive to repair the damage, but its inevitable accumulation is an important factor in the ageing process.

Thus there is nothing optimal or ideal about living on an oxygen-rich planet; it is simply the way things turned out. Oxygen is, after all, an extremely abundant element: the third most abundant in the universe, and the most abundant (47 per cent of the total) in the Earth's crust. On the other hand, the living world (the biosphere) has contrived to maintain the proportion of oxygen in the atmosphere at more or less the perfect level for aerobic (oxygen-breathing) organisms like us. If there was less than 17 per cent oxygen in the air, we would be asphyxiated. If there was more than 25 per cent, all organic matter would be highly flammable: it would combust at the slightest provocation, and wildfires would be uncontrollable. A concentration of 35 per cent oxygen would have been enough to destroy most life on Earth in global fires in the past. (NASA switched to using normal air rather than pure oxygen in their spacecrafts for this reason, after the tragic and fatal conflagration during the first Apollo tests in 1967.) So the current proportion of 21 per cent achieves a good compromise.

This constancy of the oxygen concentration in air lends support to the hypothesis that the biological and geological systems of the Earth conspire to adjust the atmosphere and environment so that they are well suited to sustain life—the so-called Gaia hypothesis. Oxygen levels *have* fluctuated since the air became oxygen rich, but not by much. In addition, today's proportion of atmospheric oxygen is large enough to support the formation of the ozone layer in the stratosphere, which protects life from the worst of the sun's harmful ultra-violet rays. Ozone is a UV-absorbing form of pure oxygen in which the atoms are joined not in pairs, as in oxygen gas, but in triplets.

How is atmospheric oxygen kept at such a steady level? It is created, as we have seen, during photosynthesis, when organisms strip oxygen from water molecules. Photosynthetic organisms include all plants and many species of bacteria. Oxygen is consumed by animals and other aerobic organisms. It is tempting to regard the steady level of oxygen as a balance between these sources and sinks in the biosphere. But there is more to it than that. The oceans act as a kind of buffer against large variations in atmospheric oxygen, since the decomposition of marine organic matter (which removes oxygen from the air) slows down if oxygen levels fall.

Oxygen is one of several vital elements that are constantly consumed and recycled by processes involving the biosphere, the Earth's rocks and volcanoes, and the oceans. These so-called *biogeochemical cycles* are linked: changes in the cycling of oxygen, carbon, nitrogen, and phosphorus are interdependent. The meshed cogs create a more or less constant environment on our planet. Changes to the turning speed of one of the cogs—for example, owing to industrial and agricultural practices that pump carbon-rich gases into the atmosphere—can upset the other cogs in ways that are hard to predict. This is why there is so much uncertainty about the likely course of global climate change caused by human activities.

Because the biogeochemical cogs are always turning, the

chemistry of the Earth is not *at equilibrium.* When a chemical process reaches equilibrium, all change ceases. The chemical constancy of our planet's environment is due not to inactivity but to perpetual change. It is the difference between a person staying on the same spot by standing still or by walking a treadmill.

This disequilibrium of the Earth's environment involves inorganic processes in sea and rock, but it is ultimately sustained by the biosphere—by life. The cogs are kept in motion mostly by the energy of the sunlight captured by photosynthetic organisms. If life ceased, the planet would gradually settle towards a static equilibrium that would be very different from today's environment.

We can see this by looking at the atmospheres of our neighbouring planets. Venus and Mars are of a similar size to Earth, and they were formed from a roughly similar mixture of elements. But their skies now contain only tiny amounts of oxygen—less than 1 per cent—and only small quantities of nitrogen. Their atmospheres are both about 95 per cent carbon dioxide, even though that of Mars is very tenuous while that of Venus is very thick. On Venus this dense blanket of the greenhouse gas raises surface temperatures to around 750 °C; on Mars the thin sheet keeps things at a frigid −50 °C or so. In either case, the absence of oxygen and the proximity of the mixture of atmospheric gases to an equilibrium mixture proclaims from afar that there is no life to be found on these worlds.

$$\boxed{3}$$

Gold

The Glorious and Accursed Element

As befits the son of a satyr, Midas was a king who loved the pleasures of this world. He ruled over Phrygia in Asia Minor, on the shores of the Aegean Sea, where he planted wonderful rose gardens.

In these fragrant surroundings Midas's gardeners one day found a dissolute old satyr called Silenus, sleeping off his drunken revels. Silenus was the foster father of the rumbustious god Dionysus, whose army was passing nearby. The satyr had become separated from the Dionysian horde and found a quiet resting spot in the gardens.

Silenus was brought before Midas, whereupon he charmed the king for five days with fantastical stories. When Midas indulgently returned his entertaining guest to Dionysus, the god of merriment rewarded the king by offering to grant him a wish. Midas asked that all he touched should be turned to gold. When he discovered that the charm worked not only on stones and ornaments but on food, drink, and even on his daughter, he soon begged Dionysus to retract the spell before he died of hunger and thirst.

It seems this was just what Dionysus expected. Laughing, he told the king to bathe in the Pactolus River, which flowed from Mount Tmolus in Anatolia. On doing so, Midas found that his golden touch had gone. But the sands of the Pactolus became rich with gold, and the precious metal could be sifted there for long afterwards.

The fable of King Midas is one of the prettiest admonitions in classical mythology against the dangerous allure of gold. Some say that Midas was King Mita of the Moschian people, who lived in Macedonia around the middle of the second millennium BC. Midas/Mita is said to have owned gold mines near Mount Bermius, which account for the fabulous wealth of the Mita Dynasty. His legendary riches, in other words, probably had a much more mundane source.

Midas escaped lightly from the curse of gold-lust. Many others who, in classical times, hungered after gold came to a sticky end. Polymnestor, a Thracian king in the time of the Trojan Wars, is one of the most tragic and the most villainous. He is entrusted by Priam of Troy to bring up Priam's son Polydorus, saving the boy from Agamemnon's murderous designs. But Polymnestor is bewitched by the gold given by Priam to cover the cost of raising and educating his son, and he kills Polydorus so that he might seize the wealth himself.

Priam's wife Hecabe discovers the deed by finding Polydorus' body washed up on the sea shore. She ensnares Polymnestor, her own son-in-law, with the bait of a promise to show him where to find a treasure horde in the ruins of Troy. When the king arrives with his own two natural sons, Hecabe stabs both children to death and claws out Polymnestor's eyes.* In his version of the legend, the Roman writer Virgil displays his horror at Polymnestor's acts while identifying the real cause of such villainy:

> He breaks all law; he murders Polydorus, and obtains gold by violence. To what wilt thou not drive mortal hearts, thou accursed hunger for gold?

To what indeed? When money is denounced as the root of all evil, we should properly understand it not as banknotes but as bright, treacherous gold. During the Renaissance, people idealized classical times as a Golden Age; but the Roman writer Propertius had no illusions about what that really meant, for with gold all doors are opened, and truth, honesty, and freedom can be overawed if the payment is large enough:

> This is indeed the Golden Age. The greatest rewards come from gold; by gold love is won; by gold is faith destroyed; by gold is justice bought; the law follows the track of gold, while modesty will soon follow it when law is gone.

---

* In truth, Polymnestor's fate—and his mendacity—vary in different versions of the legend. Another tells how his lust and avarice lead him to succumb to Agamemnon's offer of a new wife and a gold dowry if he will kill Polydorus. But on that occasion Polymnestor cannot bring himself to break the oath of protection that he made to Priam, so he kills his own son, Deiphilus, pretending the boy is Polydorus. On discovering the truth from Polymnestor's wife, Iliona, who has quite understandably abandoned her husband, Polydorus himself blinds and then slays the man he thought was his father.

It was a common theme in classical times, when men were lauded if they disdained riches and condemned if they coveted them. Socrates had a reputation for rising above the scramble for worldly wealth—he allegedly refused the gold offered by his rich pupil Aristippus. Inspired by this example, Aristippus once instructed his slaves to throw away the gold that they could not easily carry during a long journey. The inhabitants of Babytace, a town on the Tigris, are said to have buried their gold in the ground so that no one could use it. Classical writers even spoke approvingly of the lack of avarice in barbarian races such as the Scythians. The Roman general Marcus Crassus showed how foolhardy gold makes a man when he attacked the Parthians to win the yellow metal they possessed. He and his eleven legions were overwhelmed, and the Parthians, hearing of Crassus' motive, laughed as they poured molten gold into the dead general's mouth, saying: 'Thou hast thirsted for gold, therefore drink gold.'

But that was long ago. The craving for gold in more recent times is as strong as ever, scarcely dampened by the fates of the ancients. In the sixteenth century the German writer Georgius Agricola admitted that

> It is almost our daily experience to learn that, for the sake of obtaining gold and silver, doors are burst open, walls are pierced, wretched travellers are struck down by rapacious and cruel men born to theft, sacrilege, invasion, and robbery. We see thieves seized and strung up before us, sacrilegious persons burnt alive, the limbs of robbers broken on the wheel, wars waged for the same reason ... Nay, but they say that the precious metals foster all manner of vice, such as the

seduction of women, adultery, and unchastity, in short, crimes of violence against the person.

For gold the Spaniards eradicated the ancient civilization of the Peruvian Incas: Pizarro disavowed any mission to convert the heathens to Christianity, saying coldly and simply: 'I have come to take from them their gold.' For love of gold, settlers in the nineteenth-century New World met dusty deaths in the American West. In Auric Goldfinger, James Bond faces a mixture of Midas and Polymnestor, alive and well and hungry for the vaults of Fort Knox.

And the crowning irony is that gold is the most useless of metals, prized like a fashion model for its ability to look beautiful and do nothing. Unlike metals such as iron, copper, magnesium, manganese, and nickel, gold has no natural biological role. It is too soft for making tools; it is inconveniently heavy. And yet people have searched for it tirelessly, they have burrowed and blasted through the earth and sifted through mountains of gravel to claim an estimated 100,000 tonnes in the past five hundred years alone. 'Gold', says Jacob Bronowski, 'is the universal prize in all countries, in all cultures, in all ages.'

It is gold's very uselessness, its inert and detached nature, that makes it so precious. It is a supremely unreactive element and does not combine with the gases in the air. This means that the surface of gold does not tarnish, which is why it became so highly prized for making fine jewellery. Chemists indicate this lack of chemical reactivity by calling gold a 'noble' metal—a technical term that unwittingly captures all of gold's glorious history, denoting excellence and magnificence as well as an association with royalty and privilege. In the late Middle Ages, a noble was an English gold coin.

The ancient love of gold was more than skin deep. The metal's resistance to the corruption of age ensured that it continued to look lovely when other metals lost their sheen; but the attraction was not just physical. This incorruptibility was deemed by the alchemists to reflect a spiritual purity, which is why making gold was, for many of them, a religious quest more than a striving for riches. Because gold did not decay, the Chinese alchemists believed it could prolong life. Their search for the vital, youth-giving elixir was thus a kind of mission to secure the spirit of gold itself. Its yellow colour came to represent all that was profound: the dignity of humankind, the centre of the four compass directions. Yellow was the colour reserved for the Chinese emperor, like the purple of Rome.

The metals are the most familiar and recognizable of the chemical elements to non-scientists—for everyone senses the uniqueness of stolid iron, soft and ruddy copper, mercury's liquid mirror. And among these ponderous substances no element has more resonance and rich associations than gold. It is an enduring symbol of eminence and purity. The best athletes win gold metals (in a trio of metals that echoes that of the oldest coinage); the best rock bands win golden discs. A band of gold seals the wedding vows, and fifty years later the metal valorizes the most exalted anniversary of married bliss. Associations of gold sell everything from credit cards to coffee. Platinum is rarer and more expensive, and some attempts have been made to give it even grander status than gold. But it will not work, because there are no legends or myths to support it. There can be no other element than gold whose chemical characteristics have been so responsible for lodging it firmly in our cultural traditions.

Striking gold

Gold is a relatively rare metal: there is about four million times more iron than gold in the Earth's crust. Yet gold has been worked by smiths and craftspeople for seven millennia or more, whereas the Iron Age began only around 1200 BC and the use of iron did not become common until the time of the Romans. The reason is simple: because gold is unreactive, it does not readily combine with other elements in the ground but tends to occur in its 'native' elemental form. You can pick gold out of the earth if you know where to look. Iron, in contrast, combines with elements such as oxygen and sulphur to make mineral ores. The metal can be set free only by chemical reactions that drive out the other elements.*

* The earliest iron implements, found in Egyptian tombs dating from around 3500 BC, precede the Iron Age by a long margin. These artefacts are thought to have been fashioned from native iron metal found in meteorites. For centuries the Inuit people took their iron from a single large meteorite found in the Arctic snows. Iron was once more revered and more precious than gold, because it was found nowhere on Earth but came instead from the heavens. The Egyptian term for it, *baa-en-pet*, can be best translated as 'iron of heaven'.

Natural gold is almost always impure, being alloyed with silver. A natural alloy containing more than 20 per cent silver is called electrum, and was regarded by the ancients as a different metal from gold (although, as we saw earlier, metals were in any case held to differ only in degree and not in kind). The 'green gold' used in some jewellery today is electrum, containing about 27 per cent silver. It has a lemon yellow tint.

On average the Earth's crust contains two and a half parts per billion of gold: 2.5 grams of the precious metal for every thousand tonnes of rock. Gold deposits occur in places where the metal has somehow become concentrated, forming tiny crystals or flakes in the host minerals. This happens if the rocks are permeated by warm water rich in dissolved salts containing chlorine or sulphur. Gold can form soluble compounds with these substances* and so the fluids will leach it out of rocks. Then, when the briny water cools (or in some cases if it is heated further by volcanic activity), the gold precipitates and forms grains of the pure metal. Such deposits are commonly lodged in veins of quartz and pyrite. The latter mineral, a sulphide of iron, has a shiny metallic lustre and was often mistaken for gold itself: it is the infamous 'fool's gold'.

Gold veins in rocks are known as lode deposits. The principal vein is the 'mother lode', now a figure of speech as well as the name of one of the most famous deposits of the

---

* Gold sulphide is insoluble, but gold thiosulphate, compounded of gold, sulphur, and oxygen, dissolves readily enough.

Californian Gold Rush. When these lodes form at relatively low temperatures and low pressures near the surface of the Earth's crust, they can become exceptionally rich in gold. Such deposits can be found in Colorado and Nevada, and became known in the nineteenth century by the Spanish word for prosperity: *bonanza,* a term now redolent with the mythology of the Wild West.

Rain and stream water will dissolve and disperse most minerals over time, a process known as weathering. But gold resists the attrition of water, and so the grains in lode deposits are released when the host rocks are washed away. The tiny gold grains collect in the sediments of streams and rivers that pass over the lode veins, and can be washed far afield before gathering in alluvial deposits. As the gold grains tumble against the rocky stream bed, they are worn smooth and transformed into the bulbous little nuggets of prospecting folklore. The gold-rich alluvia are known as placer deposits, and they are the most abundant sources of natural gold. Since time immemorial, people found that they could extract the gold from placer deposits by sifting the fine-grained material through a mesh: the technique of panning (Fig. 3). Thus gold can be mined either from lode or from placer deposits. Digging gold out of the rock demands patience, organization, and cheap labour. The ancient Egyptians had all three: slaves carved open the mines in the Nubian desert from around 2000 BC, and the gold adorned the pharaohs and their tombs. There are no other major lodes in the Middle East; 'Nubia' derives from the Egyptian word for 'land of gold'. Judging from the description of the Egyptian mines by

A—Plank. B—Side-boards. C—Iron wire. D—Handles.

3 Gold prospecting in the sixteenth century, as depicted in Agricola's *De re metallica*

the Roman Diodorus Siculus in the first century BC, there were few more miserable jobs on earth:

> Out of these laborious mines, those appointed overseers cause the gold to be dug up by the labour of a vast multitude of people. For the Kings of Egypt condemn to these mines notorious criminals, captives taken in war, persons sometimes

falsely accused, or against whom the King is incensed . . . No care at all is taken of the bodies of these poor creatures, so that they have not a rag so much as to cover their nakedness . . . though they are sick, maimed or lame, no rest nor intermission in the least is allowed them . . . till at length, overborne with the intolerable weight of their misery, they drop down dead in the midst of their insufferable labours.

This lode gold is the stuff worked by legendary dwellers below the earth, like the dwarf who forged the treasure of the Nibelungs. Recall, however, how the Rhinegold was first found, like a placer deposit, at the bottom of a river.

Mining placer deposits is much easier: a lone prospector can do it, which was of course the impetus for the individualistic American Gold Rush. Placer deposits are widespread, and the earliest gold artefacts were made from placer gold. Pliny says in the first century AD, 'Gold is found in the world . . . as gold dust found in streams . . . there is no gold found more perfect than this, as the current polishes it thoroughly by attrition.'

The largest gold deposits ever found are those of Witwatersrand in South Africa, a vast placer deposit formed (no one is quite sure how) about 2.7 billion years ago, when all life was still single celled. An estimated 40 per cent or so of all the gold ever mined comes from Witwatersrand, and South Africa is still the world's major gold supplier. Here colonial powers fought over the elements: not gold alone, but carbon too, in the diamonds of Kimberly. Both substances are relatively inert and, until recently, without practical value. Yet both are deemed to be worth dying for and have warped the history of a continent.

Agricola retells the account by the Roman Strabo of how gold was extracted in antiquity from alluvial deposits in Colchis, the land between the Caucasus, Armenia, and the Black Sea:

> The Colchians placed the skins of animals in the pools of springs; and since many particles of gold had clung to them when they were removed, the poets invented the 'golden fleece' of the Colchians.

This was the magical hide sought by Jason and his crew of the Argo (Fig. 4). The fleece came from the winged ram Chrysomallus ('golden ram'), and hung in a sacred grove in Colchis protected by a dragon. On the one hand, this is a classic 'quest' legend. But it is also an amalgamation of various older stories. The sacred fleece was originally purple or black and was used in a sacrificial rite. It was woven into the tale of the Argonauts because they sailed to the Black Sea in search of gold, which the Colchians collected in the manner described by Strabo. From such practical considerations are legends made.

Most of the gold in vaults and in circulation today has been mined since the mid-nineteenth century, when gold production soared. Great deposits were discovered in several locations throughout the world, prompting gold rushes and providing immense riches to a fortunate few. The first lucky strike was in Russia, where gold was discovered in the Urals in the 1820s and subsequently in Siberia. By 1847 nearly

A—Spring. B—Skin. C—Argonauts.

4 Jason and the Argonauts discover the Golden Fleece. The legendary pelt probably derives from the Colchian practice of using fleeces to sift gold from river water

two-thirds of the gold produced annually in the world came from Russia; but that changed when a handful of grains was discovered in 1848 at the sawmill of Johann Sutter in California. The following year saw thousands of 'forty-niners' heading west to make their fortune. In 1851 gold was found in New South Wales in Australia, forcing the British government to end penal transportations to what had become a land of opportunity.

South African gold mining became big business in 1890, thanks to a process called cyanidation that separated the metal from its ore. Invented by Scottish chemist John Stewart MacArthur (who profited from it until his patent was declared invalid in 1896), this involved treatment of the ore slurry with cyanide to form a soluble compound of gold, followed by its precipitation using zinc. Cyanidation is still used for gold mining today. While the total amount of mined gold in the world was about 8 cubic metres in the early sixteenth century, by 1908 it had reached 1,000 cubic metres.

Even at 2.5 parts per billion, the Earth's surface still holds immense amounts of gold. But much of it will never be recovered. Only if the metal is concentrated by many hundredfold does it become economical to extract it. There is probably not much left to extract: the world's remaining mineable reserves may amount to only 15,000 tonnes, and at present 2,500 tonnes are being mined every year. We may soon have as much gold in the world as we can ever have.

There is another immense reservoir of the precious metal: the oceans. Sea water contains a tiny ten parts per trillion of gold, hundreds of times less than the crust. Even so, this implies that an awesome ten million tonnes are dispersed through the world's oceans: a prize worth over $1,500 trillion to anyone who can claim it. But it would be easier to risk the hazards of Jason's mythic quest, for it is hard to imagine how such low concentrations could ever be harvested at a profit. The German chemist Fritz Haber once believed that he could do so, and that the rewards would pay off the reparations imposed on his country after the First World War. Haber turned out to be just another of those dazzled by gold's bright charms, for he had overestimated its concentration in sea water a thousandfold.

A new and ingenious, although still rare, mining technique enlists nature's own miniature miners: rock-eating bacteria. The micro-organism *Sulfolobus acidocalderius* thrives in hot environments and metabolizes sulphur compounds. It can digest the mineral pyrite to extract sulphur, and in the process it concentrates gold within the mineral into tiny grains. This process of 'biomineralization' is used today to reclaim gold from the Youanmi deposits of Western Australia.

In 1998 a team of researchers in New Zealand demonstrated another potential form of biological mining in which gold is concentrated in plant tissues. Although various plants can collect gold in their tissues, the uptake is usually too small (no more than about four times the average concentration in the Earth's crust) to be used as an extraction method. But Robert Brooks of Massey University in Palmerston North and colleagues found that leaf mustard (*Brassica juncea*), a quick-growing plant and a so-called hyperaccumulator of metals, can amass about a hundred times more gold than normal plants. The researchers grew the plants in pots containing a synthetic 'model' gold ore, to which they added a chemical that makes gold soluble. The plants accumulated around seventeen parts per million of gold in their tissues, which is just about enough to make the process economically viable.

Agricola's *De re metallica* offers a vigorous defence of mining and happily shows nature being laid waste, trees chopped down, and rivers polluted, all in the cause of extracting metals from the earth. Wouldn't it be pleasing if the mining industry was to abandon such despoliation and instead gather its products by planting mustard?

Golden recipes

Because natural gold is never pure, ancient technologists had to develop impressive metallurgical skills to separate it from impurities such as silver. In Egypt and Mesopotamia, where these methods were devised, metalworking was sacred and metallurgists commonly laboured in compounds attached to temples. The Babylonian god Marduk was 'Lord of Gold'.

At the same time, these craftspeople concocted recipes for making artificial electrum by alloying gold and silver. It might seem strange that anyone would *want* to adulterate gold in this way, but electrum goblets were thought to have the invaluable property of detecting poison.

Parting silver from gold was not just a concern of those who took it from the ground; it was an essential skill in commerce too. Traders needed ways of assaying the purity of the gold they acquired, which could all too easily be degraded by amalgamation with silver. Probably the earliest separation method was cupellation. In its simplest form this involved heating a metal to melting point in a vessel made of a dried paste of bone ash, whereupon the impurities separated and became absorbed into the vessel walls. Cupellation was probably known by 2500 BC, although its earliest use was to separate silver from lead, not silver from gold.

Strabo (63 BC to AD 24) tells how the separation can be achieved using salt, and in the twelfth century AD the Benedictine monk Theophilus says that sulphur can be used to remove the impurities from gold. These and other methods are likely to have been devised by alchemists, for whom the distinction between purifying and making gold was often an ambiguous one. Jabir ibn Hayyan (see page 19) and the renowned sixteenth-century alchemist Basil Valentine (who was also probably fictitious) both describe purification methods. Agricola devotes many pages to recipes for purifying gold alloyed with other metals, several of which are obfuscated by unnecessary ingredients that Agricola copied without question from earlier sources.

The might of the Roman Empire came from its wealth in precious metals, not from its productivity. It fed its citizens on grain imported from the colonies, the first indication that a superpower can exist on gold alone. Much of Rome's gold came from mines at Rio Tinto in Spain, which were worked at least two thousand years ago. The ores generally yielded a blend of gold and silver with copper. The precious metals could be extracted by stirring the molten alloy with molten lead: gold and silver dissolved in the lead while copper did not. The lead alloy was then separated into its component metals by cupellation.

We have to remember that during all of this the early metallurgists had no notion that they were separating elements. All metals, they believed, were basically the same stuff (Jabir's sulphur and mercury) in various stages of maturation—which is to say, in various states of purity. Rather than being considered elements in their own right, they were regarded as among the most 'highly mixed' of substances, which is why natural philosophers interested in the elemental constituents of matter tended to avoid metals and to experiment on 'simpler' materials instead. Their similarities in behaviour prevented metals from being seen as chemically distinct.

With this in mind, those who believed in the alchemists' claims to transform base metals to gold appear far less credulous. Did the alchemists ever succeed? Of course they did! In an age when metals could be judged by little more than appearances, and when gold could be hidden in mixtures with other substances until chemical or metallurgical practices drew it forth, what else was one to believe when the magnificent yellow metal came glimmering forth from some unlikely lump of matter? Robert Boyle himself recounted (albeit cryptically, betraying his unease about the admission) a first-hand observation of an alchemical 'projection', a transmutation that created gold. Such accounts were commonplace in Boyle's time. In 1679 Johann Becher, having obtained a commission from the Dutch government to make gold from the sands of Holland, performed a successful projection in front of a governmental commission and the

Mayor of Amsterdam. His plans to scale up the process were waylaid by dissent fomented in the Dutch commission by Becher's enemies (so he claimed), and he had to flee Holland to save his life.

The quest for artificial gold has never ended. The Swedish writer August Strindberg convinced himself that he had made it alchemically in 1894; but this at least was one claim susceptible to chemical disproof, for his 'gold' turned out to be a gold-coloured compound of iron: a variety of fool's gold.

The noble and true Great Work of alchemy—the transmutation of base metals to gold—may have made many reputations and fortunes in former times, but we can be sure that it was never genuinely achieved until 1941. There is only one way to do it: by a nuclear, not a chemical, reaction. That is to say, we have to alter the very nature of the metal atoms themselves by adding fundamental particles to, or subtracting them from, the atomic nucleus. The American nuclear physicist Kenneth Bainbridge and his colleague R. Sherr fired high-energy neutrons from a nuclear reactor at atoms of mercury and managed to chip off a tiny part of the nucleus of some of the atoms, converting them to gold. I shall say more about this kind of nuclear chemical process in Chapter 5.

The money metal

'Dost thou not know the value of money; and what it serves?' asks Horace. 'It buys bread, vegetables, and a pint of wine.'

Ah, those were simple days. It is true that even in Horace's time money could also buy less prosaic things, such as a mercenary army or the services of a traitor. But those who were bought for gold would expect to be paid in gold. Money was something that weighed heavy in your hand—bright discs of precious metal, not slips of paper or an abstract concept held in the computerized accounts of a Swiss bank.

Agricola, arguing against the accusation that precious metals breed avarice and vice, recounts the benefits of a monetary system in the sixteenth century:

> When ingenious and clever men considered carefully the system of barter, which ignorant men of old employed and which even today is used by certain uncivilized and barbarous races, it appeared to them so troublesome and laborious that they invented money. Indeed, nothing more useful could have been devised, because a small amount of gold and silver is of as great value as things cumbrous and heavy; and so peoples far distant from one another can, by the use of money, trade very easily in those things which civilized life can scarcely do without.

Gold is, in other words, the lubricant of trade. It is what makes the market work. Barter relies on a coincidence in time and place of the availability of and demand for different goods. Money removes the need for this fortuitous juncture, because it holds its value and so allows the baker to sell his warm bread to the dairyman in the morning and in return to purchase his evening beer from the publican. And gold is the ideal currency metal because it carries so great a worth in so small a volume: you could keep the equivalent value of twenty cows in your purse. Once again, gold performs this duty well because of its extreme inertness: the glitter and purity of a gold coin do not diminish over time.

The first coins appear to have been minted in the seventh century BC in the Greek city state of Lydia in Asia Minor. They were made not of pure gold but of its natural silver alloy electrum, formed into discs and stamped to identify their origin. King Croesus was Lydia's last king, and until he was conquered and imprisoned by the Persians in 546 BC his wealth was legendary. Much of the Lydian gold came from the alluvial deposits of the Pactolus River, a bounty allegedly born of Midas's foolishness. Croesus replaced electrum coins with currency of pure gold and silver. During the fifth century BC the Athenians introduced the third and more lowly currency metal: bronze, an alloy of copper and tin.

The Romans were the first to discover the vicissitudes of a culture that derives its power from finance. Gold, like any other commodity, does not have an absolute value; it depends on how much of it there is around. The gold denomination of the Roman Empire was the aureus, which was worth twenty-five silver denarii. But the later emperors were prone to grotesque displays of wealth—Nero constructed a Golden House with jewel-encrusted walls. These excesses removed so much gold and silver from circulation that the coin minters were forced to add other metals to the aureus and the denarius. By the third century AD the denarius was 98 per cent copper. Naturally, a trader will not give as much for a coin that is mostly copper as for one that is pure silver, even if they are called the same thing and bear the same stamp.

The currency had, in other words, become devalued. As the purchasing power of the coins fell, more of them were needed to buy the same amount of goods: inflation increased. Those who possessed good-quality coins tended to hoard them up, trading only in alloy coins. It was the first example of the economic principle known as Gresham's law, named after Sir Thomas Gresham, founder of London's Royal Exchange: 'bad money drives out good'.

Kings, queens, and emperors were slow to learn the lesson that money is for using, not hoarding. Throughout the Middle Ages prevailing opinion held that the might of a nation was determined by the size of its coffers, and until the eighteenth century monarchs waged endless and futile wars of acquisition. Yet the health of an economy, as John Maynard Keynes was to point out, relies on money being circulated—spent and reinvested—rather than accumulated. Gold functions as an effective currency, either real or notional, only when it is put to work in its own uniquely passive and inert way.

In the nineteenth century, gold helped to bring money under control. Once the value of money became notional—a promise printed on paper or stamped on cheap metal—what was to stop a nation from increasing its wealth by simply printing more bills, sending inflation soaring? The answer was to link the value of paper money to the nation's gold reserves. In 1821 Britain officially established the gold standard. 'We have gold,' said US President Herbert Hoover (appropriately enough the translator of Agricola's treatise on metals) in 1933, 'because we cannot trust Governments.'

In countries that accepted the gold standard, currency could be exchanged at a bank for a fixed weight of gold. A £100 note in Britain would get you 22 ounces of the stuff. Britain had in fact defined its currency this way ever since 1717, when the Royal Mint was in the charge of Sir Isaac Newton.* But the gold standard could become established as the bedrock of international commerce only when other nations tied their monetary systems to gold in the same way. It was not until the 1870s that linking currency to gold reserves was standard practice throughout the world.

Adopted by all the world's major trading nations, the gold standard provided a common foundation to which the value of international currencies could be anchored. It meant that exchange rates were fixed: if a pound sterling was worth 113.0016 grains of gold, and a US dollar could be exchanged for 23.22 grains, the pound had a fixed exchange rate of $4.86. 'Currencies', says economics Nobel Laureate Robert Mundell, 'were just names for particular weights of gold.' With that assurance, you could use dollars in London and francs in New York.

But there were inherent weaknesses in fixing the value of currency this way. It made the economic fortunes of poorer

---

* Newton instigated a 'bimetallic' system in which the value of money was also linked to a fixed weight of silver. The problem with this was that the relative costs of gold and silver were apt to fluctuate as their supplies rose and fell. This meant it could become profitable to buy silver and sell it for gold, or vice versa.

nations rigidly dependent on those of richer countries on the other side of the world. 'When London sneezes,' went the saying, 'Argentina catches pneumonia.' In 1873 it was the US economy that paid the price of London's sneezes, with a depression caused by loss of confidence on the part of British investors. The same thing happened again in the 1890s. In 1896 the People's Party in the USA claimed that 'The continuance of the "present gold standard" means: Ruin; Rage; Riots; Debts; Crime; Strikes; Tramps; Poverty; Mortgages; Hard times . . .' and much misery besides. And indeed hard times were on their way in more severe measure than ever before.

If the trauma of the First World War shook the international monetary system, the Great Depression that followed strained it almost to breaking point. The result was civil unrest and the rise of extremist movements. In 1931 Britain had had enough, and it abandoned the gold standard—to the delight of John Maynard Keynes, who argued vociferously that 'gold is a barbarous relic'. Keynes was one of the main architects of the restructuring of the world financial system that was engineered at the Bretton Woods conference in New Hampshire in 1944, when international currencies were effectively anchored to a dollar standard instead. The dollar became the only currency still convertible to gold on demand—but only by treasuries and banks, not individuals. This golden anchor looked attractive after the Second World War, but was poorly designed to withstand later changes in the relative strengths of national economies. President Richard Nixon finally gave it up in 1971. No currency since then has had its value fixed by gold reserves; instead, international exchange rates are 'floating', and no government offers to redeem its currency for bright gold.

From time to time in recent decades governments have sought to revitalize the idea of tying trade and currency to the value of gold—Charles de Gaulle promoted this idea vigorously in the 1960s. But it no longer seems to be a realistic option. Without a gold standard, exchange rates fluctuate unpredictably; but trying to find stability through gold threatens to let the metal's ponderous weight drag the economy out of control.

Liquid gold

The Roman emperor Diocletian in the third century AD feared inflation coming from a different source. He worried that alchemists might undermine his currency by flooding the market with manufactured gold, and he ordered the destruction of many precious alchemical documents. But true adepts would never have been concerned with anything so vulgar as financial gain. 'As to the True Man', says Ko Hung (*c.*AD 260–340), the most famous of the Chinese alchemists, 'he makes gold because he wishes by its medicinal use to become an Immortal . . . the object is not to get rich.'

This medical aspect of Chinese alchemy distinguishes it from the metallurgically based Arabic and Western traditions, at least until the time of Paracelsus in the sixteenth century. For Chinese alchemists, gold held the key to the

Elixir, the Eastern equivalent of the Philosopher's Stone. Like the Stone, the Elixir could transmute base metals to gold; the adept Ma Hsiang reputedly paid for his wine by using the Elixir to transform all the iron vessels in the wine shop. But the Elixir, the key to gold's longevity, also made one immortal and could raise the dead. Even gold plates and drinking vessels conveyed something of their imperishability to one who dined from them. The tradition extended to India, where gold was used in ritual cleansing.

What was the Elixir? There are endless recipes for it. Some of the oldest focus not on gold but on cinnabar, the red mineral mercury sulphide. Gold deposits are rare in China and the metal is not mentioned explicitly in the alchemical or classical Chinese literature before about the fourth century BC. But later alchemists seeking this most potent of medicines were bent on imbibing the yellow metal itself. Moreover, gold made by alchemical art was considered more potent than that dug from the ground: 'Gold created by transformation, being the very essence of a variety of ingredients, is superior to natural gold,' said Ko Hung.

This tradition found its way to the West in the notion of 'potable gold' (*aurum potabile*), a medicine that, if drunk, would cure all manner of ills. It sounds perfectly mythical, for gold does not dissolve in water and it melts only when heated to over 1,000 °C. Nevertheless, medieval alchemists had at least one recipe for making an extremely potent 'water' that would consume gold metal and presumably imbibe its virtues. This recipe appears in a book written around 1310 by a Spanish alchemist who attributes the origin of the text to the works of the great Jabir ibn Hayyan. The 'dissolutive water' is made from 'vitriol of Cyprus', saltpetre (potassium nitrate), 'Jamenous Allom' (alum), and sal ammoniac (ammonium chloride), and it became known as *aqua regia*, the king of waters.

Aqua regia is basically a mixture of nitric and hydrochloric acids (Jabir's recipe would have also contained sulphuric acid), and it is one of the few chemical reagents potent enough to corrode gold. The metal forms a 'complex' in which each gold atom combines with four chloride ions; this complex is soluble in water. The disappearance of 'immortal' gold when treated with *aqua regia* must have seemed miraculous to the alchemists.

Needless to say, a mixture of concentrated nitric and hydrochloric acids is not particularly medicinal, whether it contains gold or not. But the acids could be diluted with rosemary oil without giving up their soluble gold, and the resulting potion was the fabled *aurum potabile*. At least, that is one prescription; other sources suggest making it, for example, by pouring alcohol (distilled from wine), vinegar, or urine onto hot gold or gold amalgamated with mercury. As Agricola explains, gold miners made use of *aqua regia* to separate gold from silver—albeit in a rather haphazard way, for the distinct acids were not recognized as such until much later.

Glassmakers discovered that 'soluble gold' could be used to make the most gorgeous ruby-red glass. Adding a tin compound to the solution turns the liquid a deep purple colour. The first written account of this process comes from Andreas

Cassius, a glassmaker from Potsdam, in 1685; the coloured substance became known subsequently as Purple of Cassius. The art of incorporating this 'liquid gold' into ruby glass is attributed to German glassmaker Johann Kunckel in the late seventeenth century.

To alchemists the transformation of soluble gold into a purple liquid was surely even more wonderful than the dissolution of the metal. Purple was auspicious—the colour of Imperial Rome, associated with majesty, and according to some sources the colour of the Philosopher's Stone itself. And, just as the method for making the fabled Tyrian purple dye had been lost to the West when Constantinople fell to the Turks in 1453, so too had the Romans possessed the forgotten secret for colouring glass ruby red. The Lycurgus Cup, dating from the fourth century AD and now residing in the British Museum in London, is made from glass tinted with gold. The cup looks green in reflected light, but viewed in transmitted light it appears red.

Purple of Cassius was used not just by glassmakers but by porcelain manufacturers to make a fine red glaze. But how yellow gold creates a red colour remained a mystery for almost 200 years. The ruby hue is due to tiny gold particles too small to see with the naked eye. Treating the dissolved gold complex with tin makes the gold revert back to its metallic form; but, instead of precipitating as a lump, the atoms aggregate into clusters just a few hundred millionths of a millimetre (nanometres) across. A dispersion of such tiny particles in water is called a sol, and is an example of a *colloid*: a mixture of microscopic particles of one substance in another. Colloids were named by the Scottish chemist Thomas Graham in the 1860s, after the Greek word for glue (*kolla*), which is itself a colloid.

Milk is another colloid, consisting of microscopic globules of fat dispersed in water. Because the colloidal particles are of much the same size as the wavelengths of visible light, they scatter light strongly. Milk scatters all wavelengths and so it appears white. Colloidal gold scatters mostly blue and green light, and transmits only the red. This propensity of colloids to scatter light was explained by the Anglo-Irish physicist John Tyndall in the mid-nineteenth century. At much the same time, Tyndall's colleague Michael Faraday at the Royal Institution in London found that the purple-red liquid turned blue when he added small amounts of salt. The salt allows the gold particles to aggregate into slightly larger lumps, which are big enough to scatter red light preferentially, transmitting the blue.

The tiny particles in colloidal gold were not seen directly until the early twentieth century, when the Austrian chemist Richard Adolf Zsigmondy invented the ultramicroscope, a device capable of resolving such small objects. For elucidating the nature of colloids Zsigmondy was awarded the Nobel Prize in chemistry in 1925.

Who would imagine that this red liquid holds the most precious of all metals? That was what the Danish physicist Niels Bohr counted on when Germany invaded Denmark in 1940. When war broke out, the German physicists Max von Laue and James Franck had given the precious gold medallions of their Nobel Prizes to Bohr for safekeeping. Now they were no longer safe in Copenhagen either. Germany needed gold to fund the war, and exporting the metal became a criminal offence. The medals bore the names of the recipients and Bohr risked incriminating them if he tried to smuggle the medals from the occupied territories.

Bohr's colleague, the Hungarian chemist George (György) de Hevesy, concocted a plan to keep the gold out of German hands. De Hevesy dissolved the medals in acid, creating a colloidal sol so dark it was virtually black. The liquid was kept in unmarked jars on a laboratory shelf and no one thought to wonder what they contained. After the war, the gold was recovered (all you need to do is heat the sol) and was recast into medals for the two owners.

The noblest metal

Gold owes its illustrious career to inactivity: it reacts only with great reluctance. By rights it should be chemically similar to copper, yet copper is corroded readily enough by wind and rain. Why is gold special? The answer is surprisingly subtle and was fully understood only recently.

Metals tarnish when their surface atoms react with gaseous substances in the air. Oxygen is a highly reactive element, as we saw in the previous chapter, and it combines with iron to form the ruddy oxide compound we recognize as rust. Copper reacts with oxygen and carbon dioxide to form a greenish patina of copper carbonate. Silver resists the advances of oxygen but will slowly combine with sulphur compounds in the air to form black silver sulphide.

Gold does none of these things. Yet it is not a wholly unreactive element: it will combine with other metals in alloys, and individual atoms of gold will form strong bonds with various elements. The surface of gold metal is inert, however, because of the way its electrons are distributed.

Chemical bonds result from the sharing of electrons between atoms: the electrons team up in pairs (see page 110). But electron pairing does not always bring atoms together. Rather like couples at a party, some electron pairs contribute to the congeniality of atoms while others promote fractious behaviour. The latter are called antibonding pairs, and they cause atoms to repel one another. Electrons pair up to form bonding pairs if at all possible; but, if there are too many of them, they form antibonding pairs too. If the number of antibonding pairs equals the number of bonding pairs, the atoms have no inclination to stick together.

An atom or a molecule trying to stick to the surface of gold finds that the electrons of the gold atoms are disposed to forming antibonding pairs as well as bonding pairs. Jens Nørskov and B. Hammer of the Technical University of Denmark in Lyngby discovered this in 1995 when they performed sophisticated calculations to find the electron energy states on the surface of gold and various other metals. Both copper and gold surfaces are prone to engaging in antibonding with foreign atoms, and those atoms conclude that they are better off sticking to one another than trying to forge links with the metals. Copper is reasonably inert—its slowness to react is one reason why copper alloys also make good coinage metals. But gold is even more 'noble', and continues to shine brightly when other, lesser metals have succumbed to the dulling march of time.

If there is a lesson in all of this, it is that there is nothing obvious about the properties of even the most ancient and familiar of elements. The high priests of ancient Egypt knew that gold was special; it has taken six millennia to appreciate why.

The Eightfold Path

Organizing the Elements

One summer's day, longer ago than seems reasonable, I wrote an essay about niobium. I was sitting my chemistry finals; but even so, I am rather astonished that I managed to fill several pages with an account of this single obscure element. Goodness knows what I found to say about it.

Yet perhaps it is not as surprising as all that. True, I could not possibly have memorized the quirks and foibles of all ninety-two elements up to uranium in the Periodic Table, chemistry's group portrait of the building blocks of matter. Even now, however, I can salvage a few scraps of information about niobium simply by looking at its position in this table.

I can say, for example, that it tends to form chemical bonds to five other atoms at a time, but can tolerate fewer and, at a push, more. It is a metal, probably quite a soft one, heavier than iron but lighter than lead. Many of its compounds—its combinations with other elements—will be coloured. It will be apt to form bonds to other niobium atoms—so-called metal–metal bonds. It will behave chemically in a similar manner to the element vanadium, but will be more similar still to tantalum.

I do not think this short paragraph would have got me the marks I was hoping for, but it is better than nothing. And it does not rely on my knowing anything about niobium *per se*— I can deduce it all from a knowledge of where the element sits in the Periodic Table, along with an appreciation of the general features and trends that the table displays. The table is not just a way of arranging the elements into a compact format; it is a cipher, filled with information about what each element is like, how it behaves and so on.

When the Russian scientist Dmitri Ivanovich Mendeleyev devised the Periodic Table in 1869, he was able to use it for much more impressive feats of deduction. He correctly predicted elements that had not yet been discovered: not just that they existed, but what they behaved like, their densities, and their melting points.

To understand how and why this information is encrypted in the Periodic Table, we need—and not before time, you might say—to define what we mean by an element. We got a pretty good working definition from Lavoisier: if you cannot break a substance down into clearly distinct and still more fundamental constituents, it stands a good chance of qualifying as an element. But the problem with this definition is that it depends on how good a chemist you are, or ultimately on the capabilities of your contemporaneous chemical technology.

For example, Lavoisier listed as elements 'lime' and 'magnesia'. But neither of these qualifies: lime is calcium oxide, a compound of calcium and oxygen, and magnesia is magnesium oxide. Both calcium and magnesium were first isolated in more or less pure form by the English chemist Humphry Davy in 1808, using the technique of electrolysis—splitting compounds with electricity. The metals' avidity for oxygen is too great for them to be parted by the chemical reactions available to Lavoisier, but electricity will do the job. Davy also found the elements sodium and potassium this way in 1807.*

So how can we know that today's elements are not just extremely intimate compounds waiting to be split? And for that matter, if elements are meant to be fundamental and irreducible substances, how is it that gold was made from mercury in 1941 (see page 67), or that *The Times* of 12 September 1933 was able to announce a startling new discovery: 'Transformation of Elements'?

It is time to dissect the atom.

Small worlds

Aristotle was perfectly at liberty to be sceptical about atoms, because the arguments for and against were all philosophical. Somewhat remarkably, the same was true even at

---

* British writer Edmund Clerihew Bentley immortalized the discovery in a little rhyme, which he is said to have composed in a chemistry class:

> Sir Humphrey Davy
> Abominated gravy.
> He lived in Odium
> Of having discovered sodium.

the end of the nineteenth century, when several distinguished scientists shared Aristotle's view. Wilhelm Ostwald, a German physical chemist who won the Nobel Prize in 1909, typified the conviction of many scientists that atomism was merely a convenient hypothesis and not to be taken too literally.

All this changed in 1908 when the French physicist Jean Perrin showed that the dancing motions of tiny grains suspended in water were consistent with Albert Einstein's idea that they were being struck by particles too small to see: molecules of water, composed of atoms of hydrogen and oxygen. Even Ostwald was persuaded: atoms are real.

Some would hardly have suspected otherwise. Once John Dalton, a diffident Manchester Quaker, had taken to drawing pictures of atoms in 1800, it was tempting to take them for granted. Dalton had every confidence in the 'solid, massy, hard, impenetrable, movable particles' that Isaac Newton envisaged over a hundred years earlier, and he imagined them as eternal, unchangeable bodies, however inaccessible to the human eye. Dalton appreciated the kinship of his idea with that of Democritus, and so he borrowed the Greek philosopher's term: *atomos* became 'atom'. His drawings depicted circular particles embellished with dots, lines, shading, or other symbols to distinguish different elements, which combined in fixed ratios to make 'compound particles' (which we would now call molecules) (Fig. 5).

What were these atoms made of? Dalton did not know, nor did he regard the question as particularly important. All that mattered were the *weights* of atoms, which he assumed to be

Dalton's Atomic Symbols

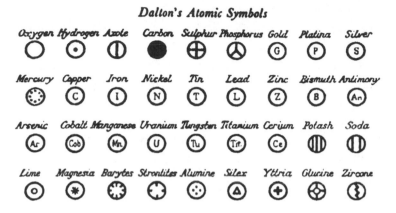

5 John Dalton's schematic depictions of atoms encouraged the view that they were small, dense, spherical particles

identical for atoms of the same element but to differ for different elements. It was known, for example, that hydrogen combined with eight times its weight of oxygen to make water. Since Dalton assumed that atoms of hydrogen and of oxygen united one to one in water, this implied that the atomic weight of oxygen relative to that of hydrogen is 8.

Hydrogen is the lightest element, so it provides a convenient unit for measuring the *relative* atomic weights of the other elements. Unfortunately, Dalton's picture of the water molecule was wrong: it contains *two* hydrogen atoms united with one of oxygen. This means that the true relative atomic weight of oxygen is 16. Errors like this meant that Dalton's list of atomic weights was a mixture of right and wrong. But no matter; later, more careful analytical chemists would correct the mistakes. (One of the most careful was the Swedish chemist Jons Jacob Berzelius, who by 1818 had deduced an improved list of atomic weights for forty-five of the forty-nine elements then known.) The important point was that the notion of atoms had been given a concrete expression, and that this helped to make sense of chemists' analyses of the composition of matter.

In short, Dalton's atomic theory allowed chemistry to become an exact science. The importance of making numerically precise measurements of chemical processes had been clear enough to Cavendish, Priestley, Lavoisier, and their contemporaries; but, without an underlying theory of the elements, these numbers were simply codifications of empirical observations. They were like measurements of the depth of a river or the number of ants in a colony—they did not reveal anything about the fundamental constitution of the system. For Lavoisier, questions about the invisible particles of matter were irrelevant to chemistry's aims.

But, if atoms were little balls that always united in the same simple ratios to make 'compound particles', this explained why chemical reactions between elements always took place in constant and simple proportions. It was why, for example, a certain mass of mercury always combined, during calcination, with another fixed mass of oxygen. French chemist Louis Joseph Proust enshrined this principle in his Law of Definite Proportions in 1788. Not that everyone agreed— methods of chemical analysis were far from perfect in those days, and so the relative amounts of each element in a particular compound might seem to vary from one experiment to another.

Dalton presented his atomic theory in his book *A New System of Chemical Philosophy*, the first and crucial part of which was published in 1808. His pictures of atoms and molecules provide a unification of the micro-world and the macro-world of chemistry: they show at once what we can observe (for example, hydrogen and oxygen combining to make water) and what we cannot: the union of real, tangible atoms. Historian of chemistry William Brock says that Dalton's symbols 'encouraged people to acquire a faith in the reality of chemical atoms and enabled chemists to visualize relatively complex chemical reactions . . . Between them, Lavoisier and Dalton completed a revolution in the language of chemistry.'

Alas, it was not all so simple. For one thing, these hiero-glyphic symbols were a huge inconvenience to typesetters, who must have breathed a sigh of relief when Berzelius (1779–1848) proposed several years later that they be replaced with an alphabetic notation for the elements. Berzelius had the eminently sensible notion that one could represent each element by the first letter of its name; or, in cases where this did not uniquely distinguish it from others, by two letters. Thus hydrogen becomes H, oxygen O, and carbon C. Cobalt is distinguished from carbon by the desig-nation Co. One might imagine that copper should have first claim to this symbol, but Berzelius was keen to insist that Latin names should be used for those elements that pos-sessed them: so copper becomes Cu (*cuprum*), gold is Au (*aurum*), and iron is Fe (*ferrum*).

Berzelius proposed that, if elements combine to form compounds with other than a one-to-one ratio of atoms, the multiplicities be denoted by superscripted numbers, later transmuted to subscripts. So the two-to-one ratio of hydrogen and oxygen in water is denoted H_2O.

This, then, is chemistry's language—its way of depicting the elements and their combinations. It is a much more transparent scheme than Dalton's. Predictably, perhaps, Dalton did not agree, saying that 'Berzelius's symbols are horrifying' and worrying that they served equally to 'perplex the adepts of science, discourage the learner, as well as to cloud the beauty and simplicity of the Atomic Theory'.

He had a point. Dalton's atomic symbols may have been highly schematic, but they were visually suggestive, reminding the reader that they refer to small, ball-like particles. Berzelius's symbols have none of this mnemonic force. Most chemists of the nineteenth century came to regard the *chemical formulae* of the compounds they studied—inscriptions such as C_6H_6 (benzene) or C_2H_6O (dimethyl ether)—as a way of abbreviating the results of elemental analyses, not as representations of any atomic model of matter. Benzene simply has six parts carbon and six parts hydrogen, rather like a cake mix.* Many chemists were inclined to disregard the question of what this implied for the way atoms were joined together. The formula H_2O does not trouble us with the issue of whether the atoms sit in the order HHO, HOH,

---

* You might ask why, if carbon and hydrogen are present in equal proportions in benzene, we don't just write the formula as CH. One answer is that the 6:6 ratio becomes evident when we consider the formulae of compounds derived from benzene, such as phenol (C_6H_6O). The 'C_6' element seemed to be a coherent unit onto which other atoms can be grafted.

or a triangle with an atom at each corner. If we were representing the compound using Dalton's atoms, on the other hand, we would be more inclined to ponder on how they are arranged. Thus the question of molecular *shape* did not occupy chemists very much until the middle of the nineteenth century; and, as I indicated earlier, even by the century's end there were those who felt it was fruitless to worry about what atoms were or how they were arranged.

Primal matter

As well as helping to give atoms a graspable reality, Dalton's theory advanced another idea. It shows the elements as distinct things, as different as a red billiard ball is from the black. What distinguishes the elements, though, is not colour but weight. Of course, making crude and speculative drawings of atoms proves nothing, but it encourages the belief that elements are not transmutable but irrevocably different from one another.

Are they, though? Some chemists, including the eminent Michael Faraday, reserved judgement on the alchemical idea of transmutation. Others were even bolder, and sought to reinvent the notion in a modern incarnation. Might elements indeed be turned one into another, if the conditions are extreme enough?

It is not only a plausible idea but a very sensible one. Dalton's atoms are distinguished only to the extent that they have different weights. Moreover, it was claimed that these weights were largely integer multiples of the weight of the hydrogen atom. So might all elements be made from hydrogen atoms, somehow squeezed together to make larger blobs?

This theory was put forward in 1815 by the chemist William Prout (1785–1850). He made no bones about the source of his inspiration: the *prote hyle* of the ancient Greek philosophers, the stuff from which all matter is derived. It was this primal substance that underpinned old beliefs about transmutation, and now Prout was apparently suggesting that this idea was valid after all. The *prote hyle,* said Prout, is hydrogen.

Berzelius called this 'Prout's hypothesis', but he never really accepted it. Others were more favourably inclined. In the 1840s the French chemist Jean-Baptiste Dumas refined the hypothesis by noting that in fact the atomic weights of some elements were *not* integral multiples of hydrogen's. Chlorine, for example, has a relative atomic weight of 35.5. (Prout had rather fudged some of his numbers to cope with these inconsistencies.) Dumas wondered whether the fundamental building blocks of atoms might be some smaller subdivision of the hydrogen atom: a half, say, or a quarter.* This basic substance became known as 'protyle'.

But no one had ever subdivided a hydrogen atom, nor had they convincingly transmuted one element into another. So

---

* Dumas was so perplexed by weight measurements of compounds that suggested atoms might be subdivided that he exclaimed in 1837: 'if I were master I would efface the word atom from science.'

why credit this untested idea? In the 1870s, the astronomer Joseph Norman Lockyer suspected that it was simply a question of finding the right conditions. Lockyer proposed that to transmute elements you needed the fiery furnace of a star.

In 1869 Lockyer discovered a new element, and it was one that had never been seen on Earth. The astronomer identified it from its imprint in the light emitted by the sun. Atoms absorb light at precise and particular wavelengths. This means that the spectrum of sunlight—the light spread out into its different colours by a prism—is interrupted by very narrow dark bands like a bar code where elements in the sun's atmosphere have absorbed the light. Lockyer saw an absorption line that corresponded to none of those measured in the laboratory for the known elements, and he concluded that it must be due to a new, hitherto unseen, substance. The French astronomer Pierre Janssen saw the same thing at the same time from his Paris observatory. The new element became named helium, after *helios*, the Greek word for the sun.

Helium is the lightest of the so-called noble gases, which are extremely unreactive elements. This is why they had not been seen earlier, for they do exist on Earth. Terrestrial helium was first found twenty-seven years after Lockyer and Janssen saw it in the sun.

Lockyer's studies of the solar spectrum revealed to him that the sun is a miasma of chemical elements. Where did they come from? In 1873 Lockyer developed the theory, later expounded in his *Chemistry of the Sun* (1887), that in the hottest (blue-white) stars the stellar matter is broken apart into the constituents of atoms themselves: *subatomic particles*, the protyle discussed by Dumas. Then, as the stars cooled, these particles combined to form regular elements— including some, like helium, not (then) known on Earth.

Lockyer thought that stars began as loose aggregates of gas and dust, replete with all manner of elements. As this material gathers more tightly under gravity's pull, it heats up until it becomes hot enough to break apart atoms into protyle. Then, while still contracting, the star cools through yellow- and red-hot, and the protyle condenses into progressively heavier elements. Thus there was a stellar *evolution* of elements, echoing Darwin's evolution of species.

This theory was presented in the journal *Nature*, which Lockyer founded, in 1914. But by that time, what atoms were made of and whether they were divisible and transmutable were questions accessible to experiments on Earth. These experiments showed that the enthusiasts of 'protyle'—Prout, Dumas, Lockyer—had hit on a kind of truth.

Inside the atom

When Ernest Rutherford (1871–1937) decided to anatomize the atom, he chose gold for the same reason that medieval artists used it to decorate their altarpieces: it can be hammered into very, very thin sheets of almost gossamer transparency. This, concluded the New Zealand-born physicist, meant that he could study a sample only a few atoms thick. That was important, for Rutherford wanted to find out what was *in* the atom. So he needed a thin section, for much the same reason that the microscopist pares off a thin sliver of tissue for investigation. He needed to see through it.

'I was brought up to look at the atom as a nice hard fellow, red or grey in colour, according to taste', Rutherford once said. But in 1907 he found that atoms were not so hard at all. They were mostly empty space. Working at Manchester University in England, Rutherford and his students Hans Geiger and Ernest Mardsen fired alpha particles from radio-active elements at thin gold foil, and found that the particles could 'see' right through this ponderous element. Mostly they passed through the foil with scarcely any deflection from their course. (Geiger helped to invent the instrument that detected the alpha particles, which he later developed into the Geiger counter.)

Of course, you would expect a bullet to fly right through gold leaf. But an alpha particle is a bullet much lighter than a single atom of gold. Rutherford deduced in 1908 that it is essentially an electrically charged helium atom: a helium *ion*. Helium has an atomic weight of 4; that of gold is 197. No matter how thin the foil is, alpha particles will not get through if atoms are like Rutherford was told they are.

But the surprise of finding alpha particles passing through gold foil was nothing compared with what Marsden found subsequently. A small number of alpha particles did not pass through at all, but bounced right back. By now accustomed to the idea that atoms are tenuous things, the researchers had drastically to revise their ideas. 'It was quite the most incredible event that has ever happened to me in my life',

Rutherford later recalled of seeing Marsden's findings. 'It was almost as incredible as if you fired a 15-inch shell at a piece of tissue paper and it came back and hit you.'

Rutherford's team had discovered the nucleus. Atoms, he concluded, are mostly empty space—but with an incredibly dense central kernel, where virtually all the mass resides. This nucleus, about 10,000 times smaller than the width of the atom itself, must be positively charged because of the way that it repels positively charged alpha particles. Surrounding it, said Rutherford, was a cloud of 'opposite electricity equal in amount'.

The Danish physicist Niels Bohr (1885–1962) turned this vague description of the atom into something more precise and conceptually appealing. Physicists had known for several years that atoms contain electrons—negatively charged sub-atomic particles discovered by Joseph John Thomson in 1897. As a young student, Bohr went to Cambridge in 1911 to work with Thomson, but he found the English physicist unreceptive and switched, as soon as he could, to Rutherford's laboratory in Manchester. In 1912 he devised a model of the atom that he published the following year and which won him the Nobel Prize in 1922.

In truth, Bohr's atom, in which electrons orbit around a dense nucleus like planets around the sun, had already been largely envisaged by Rutherford. Bohr's crucial contribution was to show how this arrangement could be stable, since, according to conventional physics, the orbiting electrons should emit light as they circulate. This means they lose energy, so that they eventually spiral into the nucleus. To get around this difficulty, Bohr had to invoke the new ideas of quantum theory, which stemmed from the work of Einstein and Max Planck at the beginning of the century.

Clearly, Bohr's atom is a long way from Dalton's. No longer is it an indivisible lump; it is made from subatomic particles—the electrons and the nucleus—and is mostly just space. The 'size' of the atom is defined not by hard boundaries but by how far the electron orbits reach.

And what of the nucleus? Rutherford proposed that it is made up of subatomic particles bearing a positive charge. He asserted that hydrogen, the lightest atom, contains just one of these particles, which he called a proton—the final incarnation of the *prote hyle* or protyle. Helium nuclei (that is, alpha particles) have twice the positive charge of hydrogen nuclei, and so, said Rutherford, they contain two protons. Here is Prout's hypothesis vindicated: since their nuclei are conglomerates of protons, all elements are, in a sense, made from hydrogen.

But that cannot be all, Rutherford realized. A helium nucleus may have twice the charge of a hydrogen nucleus but it has four times the mass. He therefore suggested that nuclei also contain particles that have the same mass as protons but no electrical charge. Rutherford's student James Chadwick discovered this neutral particle in 1932, and called it the neutron.

In the Bohr atom, as it is commonly now depicted, electrons—which have a mass just 0.00055 times that of the proton, but an equal and opposite electric charge—orbit around a nucleus of protons and neutrons, packed together with an awesome density. If matter were uniformly as dense as the nucleus rather than containing so much empty space, a thimbleful would weigh about a billion tonnes.*

This solar-system atom is so intuitively pleasing, such a neat schematization of the atom's anatomy, that it has become one of science's universal icons (Fig. 6). It is instantly recognized as a symbol of nuclear energy and is still used by the International Atomic Energy Authority. Science needs icons like this to lodge its ideas within the public consciousness.

Which is all very well, but the Bohr atom is wrong. The picture of a dense nucleus surrounded by electrons is accurate enough, but they do not follow nice elliptical orbits like those of the planets. Venus and Mars follow Newton's laws, but electrons are governed by the laws of quantum mechanics. For one thing, this makes them more fuzzy. We cannot ever pinpoint the location of an electron in an atom, even in principle; all we can do is calculate the probability of its

6 Niels Bohr's 'solar-system' model of the atom is a universally recognized symbol, still used even today to designate all things atomic—even though it is wrong. *a.* A departmental logo at the University of Chicago. *b.* The symbol of the International Atomic Energy Authority

* Matter is indeed packed to this density in neutron stars, which have collapsed under their own gravity to squeeze individual atoms out of existence.

being in a particular place at any time. This smeared-out view of the electron is a consequence of the way that very small objects display wavelike properties as well as being particle-like.

So it is better to regard the electrons as forming a kind of cloud, like bees buzzing around a hive—but moving too rapidly to see distinctly. What is more, the clouds do not adopt disc-like shapes like the rings of Saturn, as a solar-system analogy might imply. They have a range of different shapes, depending on the energies of the electrons that comprise them. Some clouds are spherical; others have dumbbell shapes or many-lobed features, centred on the nucleus. These clouds are called *orbitals*.

It was not until the advent of the quantum atom that chemists were able to understand their most abiding mystery: why elements have the properties they do. Why is helium so inert and sodium so reactive? Why do hydrogen atoms come in pairs in hydrogen gas, while carbon atoms join to four others in diamond?

These propensities are, as I indicated at the outset, largely codified in the Periodic Table. We shall shortly see that the quantum atom provides an explanation for the Periodic Table. But where did this table come from in the first place?

Patterns and affinities

One of the earliest attempts to popularize chemistry, and still one of its most enjoyable histories, is Bernard Jaffe's

Crucibles: The Story of Chemistry. It was first published in 1930, and it tells how the discipline evolved by recounting the lives of some of chemistry's most colourful characters. But do not read Jaffe if you are after an accurate historical perspective. Determined to weave a good yarn, he enthusiastically accepts every one of the popular myths, and presents each insight as a stroke of genius arriving after a furious struggle to find the truth. Jaffe's chemists seem to be on an urgent mission to establish the positivistic interpretation of science.

And so here is Jaffe's Dmitri Mendeleyev, eccentric and shock-haired 'prophet of chemistry', a 'Tartar who would not cut his hair even to please his majesty Czar Alexander III'. He is a 'dreamer and a philosopher', and the question of whether some order could be found among the profusion of elements was one that 'haunted his dreams'.

To be fair, Jaffe has for once several arguments on his side. Mendeleyev (1834–1907) was a colourful figure, no doubt about it. A descendant of Cossacks, born in the far reaches of Siberia ('of a family of heroic pioneers'), Mendeleyev clearly left an exotic impression on Sir William Ramsay when they met in 1884. Ramsay, who later discovered most of the noble gases, encountered Mendeleyev at a gathering in London and concluded that 'he is a nice sort of fellow . . . I suppose he is a Kalmuck or one of those outlandish creatures.'

Mendeleyev himself sought to foster this image of a dreamy visionary, later recalling how he finally discovered the more or less correct form of the table: 'I saw in a dream a table where all the elements fell into place as required. Awakening, I immediately wrote it down on a piece of paper.'

It is a pretty picture, and surprisingly Jaffe misses the trick of repeating it. It is not at all unlikely, indeed, that Mendeleyev drowsed while pondering the problem, or that his semi-conscious state was populated by patterns of elements. He had been working on ordering schemes for three days and nights before he found the right one, purportedly shuffling cards bearing their symbols with the temperamental obsessiveness of a manic depressive. 'It's all formed in my head,' he told a visiting friend on the eve of the breakthrough.

But the inspiration of dreams was a favourite notion of nineteenth-century chemists, practitioners of the great Romantic science. August Friedrich von Kekulé claimed that he deduced the ringlike structure of the benzene molecule this way in 1865. Scientific insights can no doubt arrive in such moments of unguarded reflection, but placing too much emphasis on or credence in them threatens to obscure the other faithful guide to discovery: prior work.*

Neither Kekulé nor Mendeleyev were the first to tackle their respective problems, and others had already glimpsed the probable solution. Mendeleyev's Periodic Table was a profound and brilliant contribution to science, but it was not the first tabulation of the elements, nor the first to highlight recurring patterns in their behaviour. When the time is ripe for advances like this, it is common for them to crystallize

---

* Some historians of chemistry entirely disregard Mendeleyev's 'dream'. Indeed, it has even been questioned whether his account of solitaire-style manipulations of cards to arrive at the correct arrangement of the elements has any foundation.

independently and almost simultaneously in more than one mind. Darwinian theory would be called Wallacian theory if Alfred Russell Wallace had rushed into print with his ideas instead of sending them to his friend Charles Darwin and agreeing to concurrent publication. And it might be the German chemist Julius Lothar Meyer, not Mendeleyev, who is immortalized by the Periodic Table if his own, similar version had been published in 1868, when he drew it up, rather than in 1870. The periodic kingdom of the elements was, by the end of the 1860s, an inevitable discovery.

Ever since Lavoisier had published his list of thirty-three elements in 1789, chemists had been seeking for ways to order and classify them. Lavoisier divided the elements into gases, non-metals, metals, and 'earths' (which included the compounds lime and magnesia). In 1829, by which time the list had expanded somewhat, Johann Wolfgang Döbereiner in Germany noticed that many elements could be grouped into threes ('triads') whose members displayed similar chemical properties. For example, lithium, sodium, and potassium comprised a triad of soft, highly reactive metals. Then there was chlorine, bromine, and iodine: pungent, poisonous, and coloured gases. These triads had their own internal logic: the atomic weight of the second member was roughly the average of the first and third.

By 1843 the German chemist Leopold Gmelin had identified ten triads as well as three groups of four elements (tetrads) and a group of five. Jean Baptiste Dumas recognized relationships between certain groups of metals in 1857. Elements, it seemed, come in families. When scientists see structure, they suspect some underlying reason for it—some ordering principle. But, to understand what brings order to the elements, they needed a scheme that embraced them all, not just a collection of occasional affinities.

One of the key components of such a scheme was provided in 1860 by the Italian chemist Stanislao Cannizzaro, who announced at an international chemical conference in Karlsruhe that the work of his compatriot Amedeo Avogadro provided an improved list of the atomic weights of the elements. This list allowed an accurate ranking of the elements by weight, from the lightest (hydrogen) to the heaviest.

Cannizzaro's weights attracted the interest of several attendees at Karlsruhe, although enthusiasm for his crusading advocacy of Avogadro's ideas was muted, to say the least. Among those who sought after the new list was Mendeleyev. But he was not alone. Lothar Meyer heard the Italian speak too, and pocketed a copy of his pamphlet. On reading it later, he said: 'It was as though the scales fell from my eyes.'

In 1864 Meyer published a table of the elements grouped according to the ratios in which they combined with one another. Kekulé had noted in 1858 that carbon tends to unite with other atoms in a ratio of one to four. In methane, one carbon atom joins with four hydrogens; in carbon tetrachloride a carbon atom is linked to four chlorines. This introduced the concept of *valency*: the ratio of atoms it takes to 'satisfy' each element. It is as if carbon atoms have four slots for other atoms. Meyer's table summarized the valencies of the forty-nine known elements and revealed that elements with chemical similarities also share a common valency. The group lithium–sodium–potassium has a valency of one, as does the group chlorine–bromine–iodine.

Others had also noticed something afoot in the way elements share affinities. William Odling gathered elements into groups in the 1850s according to their physical and chemical characteristics. These included series such as oxygen–sulphur–selenium–tellurium and nitrogen–phosphorus–arsenic–antimony–bismuth, which later appeared in Mendeleyev's table. A scheme very close to Mendeleyev's was drawn up by Odling in 1864.

In that same year the English chemist John Newlands published a series of papers showing how, if the elements were listed in order of atomic weight, each element shared properties with those eight and sixteen places later. In other words, the properties repeated periodically every eight elements. Newlands drew an analogy with music, wherein the scale begins afresh every eight notes. He presented this idea in a talk to the London Chemical Society in 1866, only to be greeted with derision. Other chemists regarded the 'law of octaves' as a mere coincidence, and one joked that he might as well have sought for such patterns after arranging the elements alphabetically. Only in 1887, after Mendeleyev had vindicated the 'octave' pattern, did Newlands receive belated recognition for his observation in the form of the Royal Society's Davy Medal.

Despite all of this, it is Mendeleyev whom we remember and honour for putting the elements in order—and even then, his original table of 1869 (Fig. 7) has plenty of oddities compared with the modern version (Fig. 8). Yet Mendeleyev's

ОПЫТЪ СИСТЕМЫ ЭЛЕМЕНТОВЪ.

ОСНОВАННОЙ НА ИХЪ АТОМНОМЪ ВѢСѢ И ХИМИЧЕСКОМЪ СХОДСТВѢ.

```
                        Ti = 50     Zr = 90     ? = 180.
                         V = 51     Nb = 94     Ta = 182.
                        Cr = 52     Mo = 96     W = 186.
                        Mn = 55     Rh = 104,4  Pt = 197,4.
                        Fe = 56     Rn = 104,4  Ir = 198.
                    Ni = Co = 59    Pl = 106,6  O· = 199.
       H = 1                        Cu = 63,4   Ag = 108    Hg = 200.
           Be = 9,4  Mg = 24        Zn = 65,2   Cd = 112
            B = 11    Al = 27,4  ? = 68         Ur = 116    Au = 197?
            C = 12    Si = 28    ? = 70         Sn = 118
            N = 14     P = 31    As = 75        Sb = 122    Bi = 210?
            O = 16     S = 32    Se = 79,4      Te = 128?
            F = 19    Cl = 35,6  Br = 80        I = 127
   Li = 7  Na = 23     K = 39    Rb = 85,4      Cs = 133    Tl = 204.
                      Ca = 40    Sr = 87,6      Ba = 137    Pb = 207.
                       ? = 45    Ce = 92
                     ?Er = 56    La = 94
                     ?Yt = 60    Di = 95
                     ?In = 75,6  Th = 118?
```

.Д. Менделѣевъ

7 Mendeleyev's Periodic Table from 1869, showing the atomic weights as they were then known

The modern Periodic Table of elements, showing element symbols with their atomic numbers. Category labels: **Alkali metals**, **Alkaline earth metals**, **Transition metals**, **Halogens**, **Noble gases**, **Superheavy elements**, **Lanthanides**, **Actinides**.

1	2	3	4	5	6	7	8	9	10	11	12	13	14	15	16	17	18
H 1																	He 2
Li 3	Be 4											B 5	C 6	N 7	O 8	F 9	Ne 10
Na 11	Mg 12											Al 13	Si 14	P 15	S 16	Cl 17	Ar 18
K 19	Ca 20	Sc 21	Ti 22	V 23	Cr 24	Mn 25	Fe 26	Co 27	Ni 28	Cu 29	Zn 30	Ga 31	Ge 32	As 33	Se 34	Br 35	Kr 36
Rb 37	Sr 38	Y 39	Zr 40	Nb 41	Mo 42	Tc 43	Ru 44	Rh 45	Pd 46	Ag 47	Cd 48	In 49	Sn 50	Sb 51	Te 52	I 53	Xe 54
Cs 55	Ba 56	Lu 71	Hf 72	Ta 73	W 74	Re 75	Os 76	Ir 77	Pt 78	Au 79	Hg 80	Tl 81	Pb 82	Bi 83	Po 84	At 85	Rn 86
Fr 87	Ra 88	Lr 103	Rf 104	Db 105	Sg 106	Bh 107	Hs 108	Mt 109	110	111	112	113	114	115	116	117	118

Lanthanides: La 57 | Ce 58 | Pr 59 | Nd 60 | Pm 61 | Sm 62 | Eu 63 | Gd 64 | Tb 65 | Dy 66 | Ho 67 | Er 68 | Tm 69 | Yb 70

Actinides: Ac 89 | Th 90 | Pa 91 | U 92 | Np 93 | Pu 94 | Am 95 | Cm 96 | Bk 97 | Cf 98 | Es 99 | Fm 100 | Md 101 | No 102

8 The modern Periodic Table of elements. The numbers indicate the atomic number of each element: the number of protons its nucleus contains. Some superheavy elements beyond meitnerium (Mt) have been observed but not yet named

contribution was pivotal. His insight was to see that the challenge was not so much to find order amongst the elements as to find the order that underlay the elements. The difference between these two things is evident in the way that Mendeleyev's table leaves *gaps*, some with a question mark inserted. He realized that the science of his time might not yet have discovered all the elements.

Even this is not entirely Mendeleyev's own innovation—Odling's table also had gaps for missing elements. But Mendeleyev went further. He used the periodic trends that the table embodied to predict, in some detail, the properties of the missing elements. And his belief in his structure was so firm that he was prepared to question the experimentally determined atomic weights in cases (such as the element thorium) where they seemed to conflict with his ordering scheme.

One by one, Mendeleyev's missing elements turned up. The one he called eka-aluminium, positioned 'below' aluminium (Mendeleyev's table was vertical where the modern one runs horizontally), was discovered in 1875 by the Frenchman Paul-Émile Lecoq. He named it, with the patriotic fervour characteristic of the times, gallium.* The German Clemens Winkler followed suit in 1886 by identifying Mendeleyev's eka-silicon, which he called germanium.

Lecoq, incidentally, did not know of Mendeleyev's table or

* *Gallia* is Latin for France, but *gallus* is Latin for cockerel, which in French is *le coq*. Was gallium's discoverer also indulging in a bit of self-aggrandisement?

his predictions when he discovered gallium, and he was rather put out to find that his discovery had already been anticipated. He argued that the density of the new element was actually quite different from that which Mendeleyev had predicted for eka-aluminium, so it could not possibly be the substance the Russian had foreseen. But a subsequent measurement showed that Mendeleyev's predicted density was spot on.

Mendeleyev's original table needed a fair bit of subsequent reshuffling, and the version that the Siberian chemist published in 1902 was not only larger (amongst other things, it included the noble gases helium, neon, argon, krypton, and xenon in an entirely new row) but also reordered, bringing it closer to today's version. It had by then been given several other incarnations.

The British chemist William Crookes, for example, devised a convoluted scheme for arranging the elements that captured Mendeleyev's groupings while also illustrating Crookes's conviction, shared with Lockyer, that the chemical elements 'evolved' in stars. Crookes believed that the elements appeared from a plasma of subatomic particles, like the plasmas that he could make in the laboratory by sending electrical discharges through gases. (Such plasmas are not, in fact, subatomic, although they contain ions.) This recombination of subatomic particles was, in Crookes's view, dictated by an oscillating electrical force, whose undulations produced the periodicities in Mendeleyev's table. Crookes represented his theory using a model of a two-lobed 'lemniscate' spiral, which he unveiled in 1888 (Fig. 9).

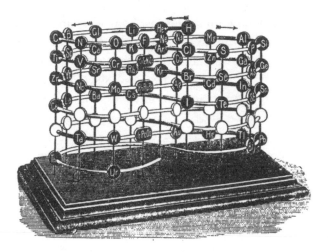

9 William Crookes's 'lemniscate spiral', an early alternative formulation of the Periodic Table

Spiral and circular forms of the Periodic Table have proved to be persistent. They were presaged by the spiral ordering of the elements essayed in 1862 by the French geologist Alexandre Émile Becuyer de Chancourtois, who found that this generated periodicities among elemental properties in vertical columns up the spiral. The Danish scientist Gustav Detlef Hinrichs also devised a kind of spiral periodic table in 1867. But none of these alternative structures has found much favour, and for the simple reason that a spiral is *too* periodic.

For after all, Mendeleyev's table has a pretty odd periodicity. The 1869 version has several unevenly sized blocks. In today's version (Fig. 8) these blocks are still there. They split apart the first row, containing just hydrogen and helium, so that hydrogen is at the extreme upper left and helium at the extreme upper right.* The next two rows contain eight elements each, in blocks of two and six. The first two of these elements are metals; the following six are (excepting aluminium) non-metals.

The fourth and fifth rows contain eighteen elements, because there is a new block of ten elements after the first two. These three rows of ten elements in the middle of the table contain nothing but metals, which are called the transition metals.

The sixth row has another block of fourteen elements interposed, and we cannot even put it in the right place without making the table inconveniently long—so it is 'projected out' below. (Some versions of the table present it as a loop emerging between lanthanum and hafnium.) This block of fourteen recurs in row seven. The members of the first of these blocks are called the lanthanide elements, and those in the second are the actinides.

That is the pattern: 2, 8, 8, 18, 18, 32 (18 + 14), 32. There is some sort of regularity in here, for sure, but it is hardly obvious. Why these numbers? What gives the Periodic Table its shape? Mendeleyev had no idea, and, until quantum theory came along, neither did anyone else.

---

* Actually, no one really knows where to put hydrogen. It is in a class of its own. Sometimes it is placed atop the alkali metals on the left, sometimes above the halogens on the right. Sometimes it is left to float freely, together with helium—which, in view of the table's trends, is probably the best solution.

Reading the table

The elements in the modern table increase in atomic weight as one goes from left to right and from top to bottom—but the weights do not advance in even steps. The progression is defined not so much by atomic weight as by atomic *number*. This is defined as the number of protons in the respective atomic nuclei.

The atomic weight of an element—the quantity chemists could weigh with scales and balances—depends on the number of protons and neutrons, which have virtually equal masses. In light nuclei, there is a roughly equal number of each; heavier atoms have an increasing preponderance of neutrons. But the number of protons is the more fundamental quantity, since it determines the positive charge on the nucleus. Until Rutherford spread the idea that atoms contain positively charged protons, there was no concept of atomic number, let alone what it might imply.

So the right way to order the elements in sequence is by atomic number, which progresses by one from each element to the next. This number tells us how many electrons the atoms of each element possess: the number of electrons is equal to the number of protons, since the protons and electrons balance one another's charge, making the atom electrically neutral.

An atom's electron count is crucial, because all chemical behaviour is determined by these particles. When atoms join together to form compounds, they do so by using their electrons as a kind of glue. There are two main ways of doing this. Some atoms like to share electrons: one of their electrons pairs up with one in another atom, making a kind of handshake. Other atoms will shed or gain electrons, becoming electrically charged ions. In the methane molecule, a carbon atom makes electronic handclasps with four hydrogen atoms. In table salt, sodium atoms donate one electron each to chlorine atoms, making the sodiums positively charged ions and the chlorines negatively charged ions (called chloride). The sodium and chloride ions then stick together by electrostatic attraction.

Either way, the bonding propensity of each element—the valency—depends on how many electrons its atoms have 'to spare'. With the exception of hydrogen (blessed with a single electron), an atom cannot use all its electrons to form bonds. Generally, only those electrons furthest from the nucleus are available for this. (Strictly speaking, it is the electrons with the highest energy that are used for bonding. These are usually the 'outermost' electrons, although that is a rather ambiguous concept in view of the strange shapes of some electron orbitals.)

When quantum theory is used to calculate how electrons are arranged around a nucleus, it shows that the electrons are grouped into shells. The first shell contains just two electrons; the next has eight, and the next eighteen. Here are the magic numbers of the Periodic Table.

Beyond the first shell, the electrons are further divided among sub-shells. The second shell has one sub-shell of two electrons and one of six. The third shell has one sub-shell of two, one of six, and one of ten. The fourth shell has sub-shells containing 2, 8, 10, and 14 electrons.

So we can start to see where the block sizes of the Periodic Table come from: they correspond to the sequential filling of shells and sub-shells by electrons as the atomic number increases. The details get a little complex, because the shells begin to overlap. For example, the first sub-shell of the fourth shell gets filled before the third sub-shell of the third shell. But in essence, new blocks of elements open up as one progresses down the rows of the Periodic Table, owing to the appearance of extra sub-shells to fill.

The periodicity comes from the fact that the filling of each shell follows much the same pattern as the one before—so the sequence of chemical properties repeats. Each element tends to form compounds that leave its atoms with a completely filled shell, either by sharing the electrons of other atoms or by adding electrons to, or subtracting them from, the outermost shell. By losing one electron to form an ion, lithium, sodium, and potassium all acquire a filled outermost shell. Carbon and silicon achieve the same thing by sharing an electron with each of four other atoms. This is the reason for the periodicity in valency seen by Meyer. The noble gases are inert because they come at the end of each row and already have a filled shell, so they do not 'need' to form bonds with other atoms to achieve this.

So that is why an element's location in the Periodic Table—its row and column—tells us a lot about its chemical behaviour. Metals fall to the left, non-metals to the right. The column number is a predictor of the valency. As a general rule, chemical reactivity declines as one progresses down the rows . . . and so on. The table is the best crib sheet a young aspiring chemist, sweating through a summertime exam, could wish for.

5

The Atom Factories

Making New Elements

So how many elements are there? I do not know, and neither does anyone else. Oh, they can tell you how many *natural* elements there are—how many we can expect to find at large in the universe. *That* series stops around uranium, element number 92.* But as to how many elements are possible—well, name a number. We have no idea what the limit might be.

Chemists and physicists have collaborated since the middle of the twentieth century to make new elements: substances never before seen on Earth. They are expanding the Periodic Table, step by painful step, into uncharted realms where it becomes increasingly hard to predict which elements might form and how they might behave. This is the field of nuclear chemistry. Instead of shuffling elements into new

---

* Elements slightly heavier than uranium, produced by radioactive decay (see later), are found in tiny amounts in natural uranium ores. Plutonium (element 94) has also been found in nature, a product of the element-forming processes that happen in dying stars. So it is a tricky matter to put a precise number on the natural elements.

combinations—molecules and compounds—as most chemists do, nuclear chemists are coercing subatomic particles (protons and neutrons) to combine in new liaisons within atomic nuclei.

It is alchemy's goal realized at last: the transmutation of one element to another. The ancient alchemists were doomed to fail because it is simply not possible to transmute the elements using chemical energy (that is, the energy involved in the making and breaking of bonds between atoms). Everything changed, however, with the discovery of radioactivity at the end of the nineteenth century—a discovery that led to one of the most remarkable, fruitful, and fateful eras in the history of chemistry. It began in a leaky wooden shed in the School of Chemistry and Physics in Paris, which Marie Curie and her husband Pierre used as a laboratory. In one sense that story ended over the city of Hiroshima in southern Japan in 1945; but in another sense it has never really ended. We are now irrevocably in the nuclear age.

How to split the atom

Marya Sklodowska, a young Polish woman, matriculated at the prestigious Sorbonne in Paris at a time when many scientists considered it bizarre that a woman should want to enter the profession at all. She married the French professor Pierre Curie in 1895, and the Curies subsequently began to study the mysterious rays that Henri Becquerel found emanating from uranium minerals in 1896. Becquerel was in turn stimulated by the discovery made by Wilhelm Röntgen the previous year. Röntgen found that a cathode-ray tube gave off rays that made a phosphorescent screen glow.

The cathode-ray tube was a favourite instrument of late-nineteenth-century physicists. Inside this glass tube, evacuated of air, a negatively charged and heated metal plate emitted a beam—a 'cathode ray'—that could be focused and accelerated by its attraction towards a positively charged plate. J. J. Thomson showed that this beam consisted of negatively charged subatomic particles, which were christened electrons. The cathode-ray tube forms the basis of television screens, in which the beam of electrons hits a material called a phosphor and makes it glow (see page 191).

But Röntgen's mysterious rays were not cathode rays. They emanated from the glass of the tube if it was struck by the cathode rays. This also made the glass glow with fluorescent light. Röntgen's rays passed through black paper, and if he interposed his hand between the rays and the glowing screen he could see the shadow of his bones in the image on the screen. For want of a better name, he called them X-rays.

Becquerel in Paris wondered whether naturally fluorescent or phosphorescent* substances might also emit X-rays. Some mineral salts containing the element uranium, a

---

* Fluorescent substances emit light when light of a different wavelength (a different colour) is shone on them. Phosphorescent materials do the same, but go on emitting for some time even after the illumination is switched off.

very heavy metal discovered in 1789 by the German chemist Martin Klaproth, were known to be phosphorescent. Becquerel knew that a uranium salt's glow was stimulated by sunlight. Yet he was amazed to discover that photographic plates wrapped in black paper became imprinted with images of the uranium salt scattered over them when they had been kept for several days in a dark drawer. It seemed that the uranium compounds emitted yet another kind of radiation, not X-rays and not related to fluorescence.

Pierre and Marie Curie called Becquerel's radiation 'radioactivity'. They found that another heavy element, thorium, was also radioactive, and deduced that natural uranium ore (pitchblende) contained other radioactive elements, which they called polonium (after Marie's native country) and radium (because it glowed). After two years of sifting through tonnes of uranium ore, they isolated salts of these new elements. The work left both the Curies with hands badly scarred from radiation burns, and it no doubt hastened Marie's death from leukaemia in 1934. Pierre might have met the same fate had he not been tragically killed in a road accident in 1906.

Marie Curie, who became a masterful analytical chemist, was awarded the Nobel Prize for physics in 1903, along with her husband and Henri Becquerel, for their work on radioactivity. Ernest Rutherford the physicist (Fig. 10) always considered it a royal joke that his Nobel Prize, in 1908, was in chemistry. But it was a strange and novel kind of chemistry that Rutherford did.

In 1899 he identified two forms of radioactivity, which he

10 Ernest Rutherford (1871–1937) deduced the basic architecture of atoms and initiated the field of nuclear physics

called alpha and beta particles. As we saw earlier, he deduced that alpha particles are helium nuclei. Beta particles are electrons—but, strangely, they come from the atomic nucleus, which is supposed to be composed only of protons and neutrons. Before the discovery of the neutron this led Rutherford and others to believe that the nucleus contained some protons intimately bound to electrons, which neutralized their charge. This idea became redundant when Chadwick first detected the neutron in 1932; but in fact it contains a deeper truth, because beta-particle emission is caused by the transmutation ('decay') of a neutron into a proton and an electron.

In 1900 Rutherford and the English chemist Frederick Soddy, working at McGill University in Montreal, showed that radioactive thorium emits atoms of the noble gas radon. Where did this inert element come from? Rutherford and Soddy concluded that thorium was *turning into a different element* by undergoing radioactive decay.

They realized that the particles emitted by radioactive elements as they decay are in fact little bits of the atomic nuclei. By expelling them, the nucleus alters the number of protons it contains, and so it becomes the nucleus of a different element. Alpha decay carries off two protons and two neutrons (a helium nucleus), and so it converts one element to a slightly lighter element two columns 'earlier' in the Periodic Table. Beta decay transforms a neutron into an electron (which is emitted) and a proton (which stays in the nucleus)—so the atomic number increases and the element moves one column *further* across the Periodic Table. Niels

Bohr and Soddy formulated this rule, which was called the radioactive displacement law.

In 1903 Rutherford and Soddy estimated the amount of energy that was released when a radioactive nucleus decayed, and found that it was 'at least twenty-thousand times, and may be a million times, as great as the energy of any molecular change' (by which they meant any chemical reaction). There was an awesome amount of energy locked up in the nucleus. If it could be unlocked, the characteristically jovial Rutherford joked, 'some fool in a laboratory might blow up the universe unawares'. Soddy was more sober: 'The man who put his hand on the lever by which a parsimonious nature regulates so jealously the output of this store of energy would possess a weapon by which he could destroy the earth if he chose.'

This was just the start. In 1919 Rutherford found that alpha particles emitted from radium could chip protons from the nuclei of nitrogen atoms. This was something new. Radioactive elements decayed spontaneously into other elements because they were fundamentally unstable. But there was nothing unstable about nitrogen. Yet Rutherford had nevertheless managed to transmute it *artificially*. The newspapers found a catchy phrase for this feat: 'splitting the atom'.

Atoms become progressively harder to split this way as they get bigger. This is because both alpha particles and atomic nuclei are positively charged, so they repel one another. To reach the nucleus and knock a bit off, the alpha particle has to break through this repulsive barrier. The bigger a nucleus, the more protons it contains and so the greater its positive charge. Alpha particles from natural radioactive sources do not have enough energy to burst through the strong electrostatic barrier around big nuclei.

The answer was to fire the alpha particles faster. Because the particles are electrically charged, electric fields can be used to accelerate them, just as the gravitational field accelerates a falling apple. In 1929 the American physicist Ernest Lawrence at the University of California at Berkeley hit on the idea of using high-voltage plates to accelerate charged particles to high speeds. The plates were shaped to induce spiral motion in the particles, since accelerating them along a straight track would require an accelerator longer than the laboratory.* With these trajectories in mind, Lawrence called his design a cyclotron.

The outer reaches

Chadwick was nearly beaten to the discovery of the neutron by the French scientist Frédéric Joliot and his wife Irène Curie, daughter of Marie and Pierre. They followed up the observation by the German physicist Walther Bothe in the late 1920s that some light elements, such as beryllium, emitted more radiation than could be accounted for, when

---

* In the early days, high-energy physicists thought as small as everyone else. Today, the particle-physics laboratory at CERN, near Geneva, runs a ring-shaped accelerator 27 kilometres long.

bombarded with alpha particles. The Joliot-Curies found that this radiation could knock protons out of the hydrocarbon molecules in wax. They decided that the mysterious emanation must consist of gamma rays, the third form of radiation produced by radioactive decay. Gamma rays are not particles but a form of electromagnetic radiation, like light, radio waves, and X-rays. It seemed implausible to other scientists that a mere gamma ray could kick protons out of wax— this was like expecting to deflect a bowling ball with a pea-shooter.

Bothe's radiation in fact consisted of neutrons, as Chadwick realized and proved in a series of experiments frantically conducted before the truth dawned on the Joliot-Curies (or someone else).*

The neutron is a better hammer than the alpha particle for smashing nuclei. Being electrically neutral, it encounters no electrostatic barrier to penetrating the nucleus. Indeed, slow neutrons often find their way into nuclei more efficiently than fast ones, much as a slow cricket ball is easier to catch. So the discovery of the neutron, in the eyes of the veteran nuclear physicist Hans Bethe, marked a turning point in the development of nuclear physics.

The Italian physicist Enrico Fermi set out to investigate

* The Joliot-Curies got their moment of glory nonetheless, when they found in 1933 that stable light elements such as boron and aluminium could be transmuted into radioactive elements by alpha bombardment. This discovery came as a great joy to Irène's mother shortly before her death, and it earned the Joliot-Curies the 1935 Nobel Prize for chemistry. Irène, like her mother, died of leukaemia.

what happens when elements are bombarded with neutrons. Although neutrons generally knock protons or alpha particles out of the nuclei of light elements, heavy elements are not so easily knocked about. They tend to absorb and capture the neutron, seizing it using the same nuclear force that binds the subatomic constituents of the nucleus together in the first place. Subsequently, and in its own sweet time, the nucleus decays by emitting a beta particle.

Fermi realized this meant that, if uranium, the heaviest known element, was irradiated with neutrons, it might decay to form a previously unknown 'transuranic' element. Uranium has an atomic number of 92; beta decay would convert it to 'element 93', a new member of the Periodic Table.

How would you know if you have made a new element? Neutron irradiation of a small sample of uranium could be expected to produce only an extremely tiny amount of element 93, perhaps a thousand atoms or so. Because they are radioactive, such atoms should be easy to spot with a Geiger counter. But first you need to separate them from the uranium, which is radioactive too. This is why the nuclear physicists needed the help of chemists. From its beginning with the work of the Curies, nuclear chemistry or 'radiochemistry' has had to work with incredibly tiny samples of rare elements, and has required a skill at analysis—separating substances into their elemental components—that Antoine Lavoisier could never have dreamed of.

Fermi enlisted the services of Italian chemist Oscar D'Agostino. By neutron irradiation of uranium they found a new beta-emitting source, which D'Agostino showed was none of the known elements between uranium (atomic number 92) and lead (atomic number 82). In 1934 Fermi reported 'the possibility that the atomic number of the element may be greater than 92'. He was cautious in voicing his conclusions, but could not resist naming the *two* new elements he and his co-workers thought they had found: they called element 93 ausenium and element 94 hesperium.

You will not find these in the Periodic Table, however, because Fermi's team never actually found elements 93 and 94. They could not have imagined that something even more dramatic had happened to their uranium. That story was not to emerge until several years later.

The first genuine transuranic element was discovered at Berkeley, where Edwin McMillan used Lawrence's cyclotron in 1939 to bombard uranium with slow neutrons. He saw beta decay from what he predicted was element 93, and set about trying to isolate it. McMillan saw that the element sits beneath the transition metal rhenium in the Periodic Table, and so he assumed it should share some of rhenium's chemical properties. But when he and Fermi's one-time collaborator Emilio Segrè performed a chemical analysis, they found that 'eka-rhenium' (in Mendeleyev's terminology) behaved instead like a lanthanide, the series of fourteen elements that loops out of the table after lanthanum (see page 190). Disappointed, they figured that all they had found was one of these known elements.

But when chemist Philip Abelson joined McMillan in 1940, he quickly proved that eka-rhenium was indeed a new element, with properties similar to uranium. McMillan named it 'neptunium', after Neptune, the next planet out from Uranus. It was the start of a voyage into the outer reaches of the Periodic Table.*

Towards the end of that year, Glenn Seaborg, Joseph Kennedy, Edwin McMillan, and Arthur Wahl at Berkeley used a cyclotron to bombard uranium with ions of heavy hydrogen (deuterium; see page 152). They produced neptunium, which decayed by beta emission, shunting the element one place further along the Periodic Table. Subsequently the Berkeley team, supplemented by Segrè, made this new element, with atomic number 94, by firing neutrons at uranium. Wahl and Seaborg found a chemical method to separate the new element early in 1941. Following the tradition initiated by Klaproth and observed by McMillan, Seaborg named it plutonium. Pluto is the farthest planet in the solar system; he is also the Greek god of the dead.

The team wrote a paper describing their discovery, but

* This was not, however, the first synthesis of a previously unknown element. That claim belongs to technetium, element 43, which Segrè and his colleague Carlo Perrier identified in 1937. It was made in the Berkeley cyclotron by bombarding molybdenum foil with nuclei of heavy hydrogen (deuterium). It seems possible that technetium was actually made in 1925, when a German team claimed to have found a new element (they called it masurium) after irradiating the mineral columbite with an electron beam.

Another previously unknown element, astatine (element 85, the heaviest of the halogens) was made at Berkeley in 1940 by bombarding bismuth with alpha particles. Again, Segrè was among the team of chemists who showed it was a new element.

then decided to withhold it. Plutonium, they realized, was too hot for the public news during wartime.

Falling apart

In 1934 the Hungarian physicist Leo Szilard filed a patent with the British Patent Office. It was based on an idea, nothing more—an idea about how to harness nuclear energy. The Joliot-Curies had shown that bombarding nuclei with particles can induce radioactive decay artificially. And the work of Bothe and Chadwick had demonstrated that some radioactive nuclei emit neutrons. So what would happen if neutrons induced nuclear decay that led to more neutrons? The result might be a chain reaction: a self-sustaining release of nuclear energy.

It was a speculative proposal. Szilard supposed that neutrons might be better at triggering radioactive decay than alpha particles—but no one had yet shown this. And it required the identification of a substance that both captured and emitted neutrons. Moreover, to set off a chain reaction, the number of neutrons emitted would have to exceed the number captured. All the same, the possibility pointed to a dramatic, even terrifying conclusion, and it chilled Szilard to the core. If the chain reaction goes on amplifying itself, he said, 'I can produce a bomb'. He filed the patent on 12 March, the day Marie Curie died.

Four years later, the substance needed for Szilard's chain reaction was identified—in Hitler's Germany. Otto Hahn was a radiochemist working at the University of Berlin. He and his colleague Fritz Strassmann were studying the effect of bombarding uranium with neutrons, and in 1938 they found something they could not explain. Instead of the usual decay processes that chipped fragments from the nucleus, they seemed to be finding the element barium in the products. But barium has an atomic number of 56, barely more than half that of uranium. Surely uranium nuclei could not be falling in half? Could they?

Hahn confessed his perplexity that Christmas in a letter to his former colleague Lise Meitner, an Austrian physicist whose Jewish heritage had forced her to flee the Nazis for refuge in Stockholm. Meitner had begun the neutron-bombardment experiments with Hahn in 1934, before escaping from Berlin after the Anschluss in 1938. She shared his disbelief, replying:

> Your results are very startling. A reaction with slow neutrons that supposedly leads to barium! ... At the moment the assumption of such a thoroughgoing breakup seems very difficult to me, but in nuclear physics we have experienced so many surprises, that one cannot unconditionally say: it is impossible.

That Christmas Meitner was visited in Sweden by her nephew, the physicist Otto Frisch, who was another fugitive from the Nazi regime. They discussed the problem while walking in the woods, and began to accept the inevitable conclusion: the uranium nucleus was indeed dividing into two large lumps, like a droplet of water splitting in two. Back in Copenhagen in the New Year, Frisch asked a visiting American biologist what one called the process of cell division, of which the splitting of the uranium nucleus reminded him. 'Fission', he was told. And so this was the name that Meitner and Frisch gave to the phenomenon Strassmann and Hahn had observed: nuclear fission.

Philip Morrison was a young student of the American physicist Robert Oppenheimer at the time, and he recalls: 'when fission was discovered, within perhaps a week there was on the blackboard in Robert Oppenheimer's office a drawing—a very bad, an execrable drawing—of a bomb.'

Plutonium and the bomb

Why a bomb? Because fission of uranium produces not only barium and other elements, but neutrons. This was what Szilard's chain reaction needed.

But making a bomb was not so easy. Natural uranium comes in two forms, or *isotopes* (see page 150). These have the same number of protons (92) in their nuclei, but different numbers of neutrons. One isotope has 143 neutrons (uranium-235 or ^{235}U), the other has 146 (uranium-238 or ^{238}U). Only ^{235}U undergoes fission induced by the slow, low-energy neutrons that the fission products emit. So only this isotope can be used to create a runaway chain reaction. But natural uranium is mostly ^{238}U; only 1 per cent is ^{235}U. A bomb requires a 'critical mass' of only a few pounds of ^{235}U—with less than that, too many neutrons leak away for the reaction to be sustainable. But extracting even this much of the lighter isotope from natural uranium looked like an almost impossible task in 1940.

Almost, but not entirely—and that was what worried people like Szilard. He was sure that German physicists working under the Nazis would foresee the same possibility, and that they would try to build a bomb. As indeed they did, although the German atomic bomb project, led by physicist Werner Heisenberg, never got very far.* But Szilard was desperate to persuade the Americans—the only other country with the resources for such a task—to attempt to make an atomic bomb before their enemies did. He was a mere physicist, but he had an influential friend who had become much more than that: Albert Einstein.

By agreeing to write to President Roosevelt in support of Szilard's idea, Einstein unwittingly linked his name with the bomb for ever. The 1949 cover of *Time* magazine that juxtaposed Einstein's famously shaggy features against a mushroom cloud sealed in the public consciousness the notion that Einstein somehow 'invented' the bomb. In fact, this ultimate weapon was the product not of his most celebrated abstraction, $E = mc^2$, but of a prodigious feat of chemical and mechanical engineering bankrolled by the US military.

---

* Some historians think that Heisenberg may have deliberately dragged his feet so as not to deliver the bomb into Hitler's hands. Others feel he may simply have been hindered by mistakes in his calculations. We may never know for sure what Heisenberg's intentions were. The issue is explored with great ingenuity in Michael Frayn's 1998 play *Copenhagen* (London: Methuen).

But with Einstein's advocacy the Manhattan Project began, under the leadership of Oppenheimer. Named for the New York office of the Army Corps of Engineers, it was given almost a blank cheque when America entered the war after the attack on Pearl Harbor. The bomb was pursued on two fronts. One involved developing physical and chemical techniques for separating the isotopes of uranium, milligram by milligram. The other proposed using a different nuclear explosive: plutonium.

This was why the discovery of element 94 by Seaborg and colleagues was so sensitive. In 1941 the Berkeley team told the US government that one isotope of the new element, plutonium-239, could be split with slow neutrons even more efficiently than could ^{235}U. Again, a grapefruit-sized lump could make a bomb.

Creating plutonium atom by atom in the Berkeley cyclotron was no way to gather a critical mass, however. Enrico Fermi demonstrated a better means of synthesizing plutonium in 1942, when he and co-workers produced the first controlled nuclear chain reaction in a reactor at the University of Chicago. This used natural uranium fuel, which was converted to plutonium by self-sustaining neutron emission and capture. The chain reaction was held in check by rods of cadmium, which absorb neutrons, and the emitted neutrons were slowed down to fission-inducing speeds by 'moderator rods' of carbon (graphite).

Fermi's 'atomic pile' was just a prototype. For manufacturing bomb plutonium, a plant was built at the tiny village of Hanford in Washington State. And so, drip by drip, the US war machine squeezed out its uranium-235 and plutonium, while the problem of how to build an atomic bomb was tackled by the physicists, chemists, and engineers at the Los Alamos complex in New Mexico.

The rest is history—history that transformed the twentieth century, history that divides one kind of world from another. Oppenheimer, Szilard, Bohr, Fermi, and the others knew that this was indeed the significance of their quest, and for the most part they were as exhilarated by the challenge as they were dismayed by the goal. At the Trinity test of July 1945, when the first nuclear bomb was exploded in the Nevada desert, Oppenheimer recalled words from the Hindu scripture *Bhagavad Gita*: 'Now I am become death, the destroyer of worlds.' To the US military, on the other hand, this was just a bomb—albeit one powerful enough to cow the Japanese emperor into submission and end the murderous war in the Pacific.

Hiroshima was destroyed on 6 August 1945 by Little Boy, a uranium bomb brought to critical mass by firing one piece of uranium at another using a gun mechanism. This brought the pieces to critical mass quickly enough to avoid the chain reaction blowing the uranium apart before most of it had undergone fission. Fat Man, dropped on Nagasaki three days later, was a plutonium device in which the man-made element was compressed to critical mass by an implosion. Estimates of the casualties vary widely, but perhaps 300,000 people died in the two blasts and the aftermath. Hearing the news, Szilard wrote 'It is very difficult to see what wise course of action is possible from here on.'

The power of a star

Edward Teller, one of the brilliant physicists who fled Hungary before the war and a key member of the Los Alamos team, had no doubts about the right course of action. He urged the US government to pursue the idea he had discussed with Fermi in 1942: a 'superbomb' that liberated nuclear energy not by fission but by fusion. The fusion bomb creates, for a blinding instant, an artificial sun.

The energy available from fusing light elements to make heavier ones was clear to Francis Aston, working in the Cavendish Laboratory at Cambridge in 1919, when he devised a new instrument for measuring atomic weights very accurately. This device, which Aston called a mass spectrograph (we would now say 'mass spectrometer'), led to the discovery of isotopes (see page 151).

Aston found that the masses of individual isotopes were almost exactly integer multiples of the mass of the hydrogen atom. That was as expected, for Rutherford had identified the hydrogen nucleus—the proton—as the building block for all nuclei. (The neutron was not yet known, but was inferred.) But why *almost* exactly? Aston noted small but important discrepancies between his masses. For example, the helium atom weighed slightly less than four hydrogen atoms. Where had the missing mass gone?

Aston realized that it had been transformed to energy: the energy that binds the nuclear particles together. Einstein had, after all, shown that mass and energy were interconvertible.

When Aston used Einstein's iconic equation to calculate this binding energy, he found that the tiny decrease in mass implied that an enormous amount of energy must be released when hydrogen atoms fuse to make helium. 'To change the hydrogen in a glass of water into helium', he said, 'would release enough energy to drive the *Queen Mary* across the Atlantic and back at full speed.'

Aston regarded the possibility of harnessing this process of nuclear fusion as a tremendous opportunity—and an immense danger. 'We can only hope', he said, 'that [man] will not use it exclusively in blowing up his next door neighbour.'

The French physicist Jean Perrin suggested that this might be the source of the energy that has fuelled the sun, day after day, for four and a half billion years. The astronomer Arthur Eddington agreed, saying in 1920: 'What is possible in the Cavendish Laboratory may not be too difficult in the Sun.' The idea gained currency in 1929 when the American astronomer Henry Norris Russell proved that hydrogen is the main constituent of the sun.

Hydrogen is the solar fuel, and the sun 'burns' it, not by combining it chemically with oxygen, as Cavendish and Lavoisier had done in the 1770s, but by fusing its nuclei to make helium. But hydrogen nuclei are just protons, whereas helium nuclei contain neutrons too. Where do the neutral particles come from?

They are formed by a kind of reverse beta decay: a proton becomes a neutron. In order to do so, it must shed its positive charge, and this happens by the emission of a *positively* charged version of the electron: the positron, which is the antimatter sibling of the electron.*

In the first step of hydrogen fusion, two protons combine to form a *deuteron* and a positron. A deuteron is the nucleus of an isotope of hydrogen: heavy hydrogen, or deuterium. It consists of a proton and a neutron.

In the second step, a deuteron combines with a proton to form the nucleus of helium-3, which contains two protons and one neutron. Then two helium-3 nuclei combine and spit out two protons, forming helium-4 (Fig. 11). This set of nuclear reactions is responsible for 85 per cent of the hydrogen-to-helium transmutation in the sun; other fusion

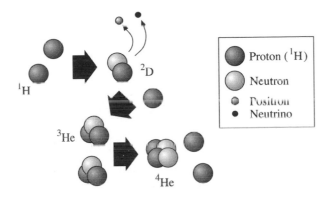

11 The fusion of hydrogen atoms in the sun creates helium-4 and releases a tremendous amount of energy

* Antimatter, predicted by British physicist Paul Dirac in 1930, self-destructs when it encounters normal matter, and the masses are converted into a burst of energetic gamma rays.

processes account for the remainder. About 600 billion kilograms of hydrogen are burnt to helium in the sun every second. As British writer Ian McEwan says, this is the 'first step along the way in the generation of multiplicity and variety of matter in the universe, including ourselves and all our thoughts'.

All it takes to trigger fusion in hydrogen are conditions extreme enough: a high enough density of hydrogen, and a temperature of about ten million degrees. Thus fusion is a *thermonuclear* process: one that sustains itself purely by the heat it generates.

In 1939 Hans Bethe showed that the conversion of hydrogen to helium can be assisted (catalysed) by small amounts of carbon. The carbon interconverts with nitrogen and oxygen in a six-step cyclical process that ends up where it began— with carbon—but that at the same time transforms hydrogen to helium. This is called the carbon cycle or CNO cycle (referring to the three elements it involves). For stars somewhat larger than the sun, this catalytic cycle provides a significant proportion of the fusion energy.

But stellar chemistry does not stop there. In 1957 the astronomers Margaret and Geoffrey Burbidge, William Fowler, and Fred Hoyle proposed a sequence of fusion reactions that get triggered at different stages in a star's life, creating increasingly heavy elements. Once a star has burnt most of its hydrogen to helium, it starts to cool. The core then begins to collapse under its own gravity, and this heats it up. The heat puffs out the outer atmosphere, which glows red. The star becomes a red giant.

As the core contracts, it gets hotter, and once it reaches about a hundred million degrees the fusion of helium atoms becomes possible. This produces carbon, oxygen, and neon. (The intervening elements beryllium, boron, nitrogen, and fluorine are less stable, and decay to other elements.)

Once the helium has been used up, the same process is repeated. The star cools, the core collapses further and heats up, and new fusion processes are ignited: carbon and oxygen are fused to make sodium, magnesium, silicon, and sulphur. Gradually, the Periodic Table emerges in this juddering, unstable furnace.

So Norman Lockyer and William Crookes (see pages 92, 106) were right in a way, if not in the details: there *is* an evolution of elements in stars. The creation of elements in stars is called nucleosynthesis, and it is responsible for the Earth and almost everything we see on it. Only hydrogen, plus some helium and a mere smattering of other light elements, are 'primordial'—products of the Big Bang. Everything else was forged in stars.*

Once a star's core temperature has reached about three billion degrees, fusion processes generate iron. And here they stop, because iron is the most stable nucleus of all. There is no energy to be gained by fusing iron nuclei. Yet heavier elements clearly do exist. They are created in the

* Well, not quite everything. The light elements lithium, beryllium, and boron are formed mostly by the break-up of heavier nuclei when hit by cosmic rays and other high-energy particles in interstellar space. This process, which whittles the nuclei down to lighter elements, is called spallation. Nucleosynthesis in stars produces very little of these three elements.

outer regions of the star, where neutrons emitted by fusion reactions are captured by nuclei to build all the elements up to bismuth (atomic number 73).

These elements are scattered throughout the universe when massive stars end their lives. When there is no fuel left to burn, the core collapses once again, and there is nothing to stop it. A shock wave from this collapse causes a rebound that fuels an enormous explosion: a supernova. The outer layers of the star are blown out into space, and the energy that is released triggers new nucleosynthesis reactions, which make the heavy elements beyond bismuth—up to uranium, and at least a little beyond.

Fermi and Teller realized in 1942 that fusion of hydrogen could release much more nuclear energy than fission of uranium. The problem was how to make the hydrogen hot and dense enough. In fact, achieving temperatures like those that drive fusion in the sun is quite impractical; but the fusion of the heavier isotopes of hydrogen—deuterium and tritium (pages 152–3)—requires less extreme conditions. This is the process exploited in the 'super-bomb'—the hydrogen bomb.

Fusion in H-bombs is ignited by a fission chain reaction of uranium or plutonium: an 'atom bomb' is used to set off the hydrogen bomb. The first hydrogen bomb test was conducted in 1952 on the Pacific atoll of Eniwetok in the Marshall Islands. Lacking the inspiration of Oppenheimer's theological preoccupations, the test was prosaically coded 'Mike'. A thousand times more destructive than Little Boy, it vaporized the island where the bomb had stood, and carved out a crater two miles wide and half a mile deep. The USA and the USSR decided over the next two decades that they needed many thousands of these devices—several times more than it would take to blow up the world.

Manfacturing elements

Thanks to the nuclear tests of the 1950s and 1960s, plutonium is now detectable in minute traces—just a few atoms—in the body of every person on Earth. It is not really dangerous in such small quantities; but plutonium is nevertheless hazardous if ingested and absorbed into bone marrow, where its alpha radiation can destroy cells or initiate cancers.

But for chemists, the hydrogen bomb tests had a happier fallout too. Scientists at the Mike test collected coral from a nearby atoll contaminated with radioactive debris, and sent it to Berkeley for analysis. There the nuclear chemists found two new elements, with atomic numbers 99 and 100. They were named after two of the century's most creative physicists: einsteinium and fermium.

There are several spaces in the Periodic Table between plutonium (element 94) and einsteinium (element 99). But by 1952 these had already been filled by scientists at Berkeley, using the cyclotron to bombard heavy nuclei with particles that, when captured, increased the nuclear mass. In 1944 Glenn Seaborg, Albert Ghiorso, and Ralph James made elements 95 and 96 this way. Kept secret until after the war, they were respectively called americium and curium.

Seaborg, Ghiorso, and others went on to make berkelium (element 97) in 1949 and californium (element 98) in 1950. The *New Yorker* wondered why they had not gone for broke, naming these two 'universitium' and 'offium' so as to reserve berkelium and californium for the next two elements. The Berkeley team responded by explaining that they did not wish to be beaten in the race by some New Yorker who could then call elements 99 and 100 'newium' and 'yorkium'.

It was not an entirely flippant comment. By the 1950s laboratories elsewhere in the world had caught on to the Berkeley technique of making elements using nuclear bombardment in particle accelerators. The Berkeley radiochemists were still leading the race when they made element 101 in 1955; and Dmitri Mendeleyev might have been either amused or perturbed to find himself immortalized, as mendelevium, in a Periodic Table that was now expanding at an alarming rate. But element 102 produced a contested finish. A group in Stockholm believed they had made it in 1957, and proposed the patriotic name nobelium, after the Swede Alfred Nobel. Their claims could not be confirmed by other element-makers, however, and element 102 was not really made until 1958 by Ghiorso and co-workers. In the same year, it was reported by a Russian team at the Joint Institute for Nuclear Research (JINR) in Dubna. No one saw fit to dispute the Swedish name on this occasion; but such unanimity was not to last.

In the 1960s and 1970s the race to make new 'superheavy' elements became more factious. Claimed sightings by one group were disputed by another, and the naming business became nationalistic and controversial. To secure a new name, putative discoverers have to win the approval of the International Union of Pure and Applied Chemistry (IUPAC), which pronounces the last word on nomenclature. No one could really argue with the choice of lawrencium for element 103, after the man who had invented the machine for element synthesis. And the Berkeley name for element 104, rutherfordium, was surely an honour due to one of the century's greatest nuclear physicists. But element 104 had been claimed five years earlier, in 1964, by the Russian team at Dubna, who wanted to call it kurchatovium after their head of nuclear research. The Americans disputed the Russian results.

Similar controversies were to plague the discovery and naming of elements 105, 106, and 107. IUPAC felt compelled to establish a working group in 1987 to assess claims to priority and to pronounce on nomenclature. By 1994, however, the question of what to call the new human-made elements was still in disarray.

Element 106 was a particularly thorny matter. It had been claimed by the Dubna team in 1974, and almost immediately after by the Berkeley team, whose evidence was more substantial. In 1993 the Americans established to IUPAC's satisfaction that theirs was the stronger claim, and the team, headed by the veteran Albert Ghiorso, proposed a name for the new element: seaborgium, after the discoverer of the first artificial element.

The trouble was that Glenn Seaborg was still alive, though

no longer really active in nuclear chemistry. IUPAC insisted that it would not do to name an element after a living person. The American Chemical Society rebelled against the IUPAC decision and approved Ghiorso's choice. In 1996 IUPAC relented, and revised all the names once again from elements 104 to 107. Rutherford was commemorated for 104, the Russians saw their efforts acknowledged with 105 (dubnium), seaborgium was accepted for 106, and 107 was called bohrium after Niels Bohr. Pity poor Frédéric and Irène Joliot-Curie, and Otto Hahn, who enjoyed the brief, posthumous glory of a place in the elemental firmament of the Periodic Table (as 'joliotium' and 'hahnium') only to be later stripped of their honours.

Searching for the island of stability

Bohrium signalled the debut of a new team of element-makers, who have dominated the field since the early 1980s. At the Institute for Heavy Ion Research (GSI) in Darmstadt, Germany, nuclear physicists perfected a new approach explored but then abandoned at Dubna. Instead of firing small, light nuclei such as alpha particles (helium nuclei) at large ones to boost the mass little by little, the GSI group fuse two medium-sized nuclei to make new superheavies (Fig. 12). For example, a lead target is bombarded with a beam of accelerated nickel or zinc ions. The former method is called 'hot fusion', since it requires that the new nuclei 'cool down' by emitting neutrons. The

12 The particle accelerator used at the Institute for Heavy Ion Research (GSI) in Darmstadt, Germany, for fusing atomic nuclei to make new superheavy elements. The GSI team have used this equipment to make all the elements from 107 to 112

latter is called 'cold fusion',* as it does not leave the new nuclei with much excess energy. The Dubna group had made fermium and rutherfordium this way in the 1970s.

Between 1981, when bohrium was made at GSI, and 1996, the German team made all the elements from 107 to 112. (Element 110 was claimed earlier, but less convincingly, at both Dubna and Berkeley.) Element 108 is called hassium, after the German state of Hesse wherein Darmstadt is situated; element 109 is named meitnerium, for Lise Meitner, who was the first to realize that uranium undergoes nuclear fission. Beyond this, the new elements have yet to be named.

As these superheavy elements get heavier, they become less stable: the nuclei sit around for progressively shorter times before undergoing radioactive decay. Plutonium-239 has a 'half-life' of 24,000 years, which means that it takes this long for half the atoms in a sample of ^{239}Pu to decay. Californium-249 (element 98) has a half-life of 350 years; mendelevium-258 (101), fifty-one days; seaborgium-266 (106) twenty-one seconds. Isotope 272 of element 111 has a fleeting existence with a half-life of 1.5 milliseconds, and that of isotope 277 of element 112, made in 1996, is less than a third of a millisecond. This is one reason why it becomes increasingly hard to make and see these superheavy elements.†

---

* Not to be confused with the 'cold fusion' of deuterium purportedly achieved by chemists in Utah in 1989 using nothing but heavy water in an electrolysis cell. This claim of cold nuclear fusion was later shown to be untenable (see page 188).

† Plutonium has a longer-lived isotope than ^{239}Pu, and it may turn out that some of the larger superheavy elements also have isotopes with greater

But nuclear scientists now realize that the stability of big nuclei does not inevitably decline as they get bigger. It can rise and fall, depending on the number of protons and neutrons the nuclei contain.

These fundamental particles arrange themselves in concentric 'shells' in nuclei, much as the electrons are arranged in shells around the nucleus (see page 110). Just as a full shell of electrons makes for a particularly stable, unreactive element (as in the noble gases), so filled shells of protons or neutrons confer stability on a nucleus. Helium, oxygen, calcium, tin, and lead all have a filled outer shell (a so-called magic number) of protons, and so their nuclei are unusually stable. It is also possible for a nucleus to have a filled shell of neutrons, and the isotope lead-208 has 'magic' numbers of both protons and neutrons—it is doubly magic.

One isotope of element 114, with 184 neutrons, is predicted to be another doubly magic nucleus, and is therefore expected to sit right in the middle of an 'island of stability' in the space of superheavy nuclei (Fig. 13). Nuclear scientists suspect that it may have a half-life of as much as several years.*

longevity than those indicated here. Nevertheless, the trend is clear enough.

* All of this remains somewhat uncertain. It now appears that the enhanced stability around element 114 may not correspond to an island at all, but could be connected to the peninsula of lighter stable elements by a narrow isthmus. Estimates vary for the lifetime of 'doubly magic' element 114, with 184 neutrons. And in fact some theorists believe that 114 might not be a 'magic' number of protons at all, but that the next magic number is 126. At the time of writing, the picture is changing rapidly.

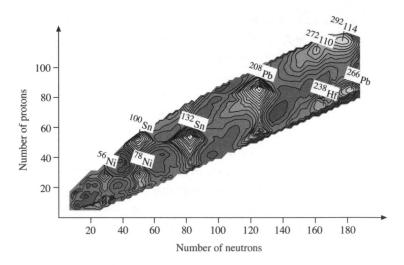

13 The isotope of element 114 with 184 neutrons is predicted to be especially stable, since it has 'magic numbers' of both protons and neutrons in its nuclei. This element may sit atop an 'island of stability' in the sea of possible combinations of subatomic nuclear particles. Other islands, here picked out in contours whose 'height' denotes the degree of stability, occur for lighter elements such as some isotopes of lead and tin

Element 114 has thus become a kind of Holy Grail for element-makers. If it turns out to be stable, this would show that these researchers are not necessarily doomed to search for increasingly fleeting glimpses of ever heavier and less stable new elements. There might be undiscovered elements out there that you can (in principle, at least) hold in your hand.

In 1999 a collaboration between the Dubna team and scientists at the Lawrence Livermore National Laboratory in California, led by Russian physicist Yuri Oganessian, cautiously announced the possible sighting of element 114. They created it by bombarding plutonium-244 with calcium-48 ions accelerated in a cyclotron. A single atom of element 114 appeared to last for about thirty seconds before decaying to element 112. This is not quite the longevity hoped for, but it is a lot longer than a third of a millisecond. And, after all, the putative isotope formed was not doubly magic, but contained just 175 neutrons, nine short of a full shell—so there should be room for improvement. Berkeley's veteran nuclear chemist Albert Ghiorso greeted the news by saying, 'This is the most exciting event in our lives.'

The Dubna researchers tried unsuccessfully for many months to repeat their apparent synthesis of element 114. Eventually, however, their persistence was rewarded with the sighting of a different isotope of element 114, with 174 neutrons and a lifetime of a few seconds. This time the researchers saw two separate decay events, making the sighting much more secure. Encouraged by this success, they changed the target material to californium-248 and manufactured element 116, which decayed by alpha emission to element 114.

But how does one get to the centre of the putative island of stability, where the doubly magic isotope of element 114 resides? This means sticking more neutrons onto the nucleus, and no one yet knows how to do that.

Single-atom chemistry

There are undoubtedly more elements on the way, as little by little the Periodic Table is extended into uncharted waters. And as this happens, we will learn about how these new elements behave. In 1997 an international team that included scientists from GSI, Berkeley, and Dubna was able to deduce that element 106 (seaborgium) has chemical properties similar to molybdenum and tungsten. In a sense this might have been expected, since seaborgium sits below these elements in the Periodic Table. But in fact the result was a surprise, because the chemical behaviour of the pre-ceeding superheavy elements 104 and 105 is distorted by the effects of relativity on the electrons surrounding the immense nuclei.

According to Einstein's theory of relativity, objects gain mass when they travel at speeds close to the speed of light. In very heavy elements the electrons are drawn into such tight orbits around the highly charged nuclei that they attain speeds big enough to experience such 'relativistic' changes of mass. This shifts the arrangement of electrons, and thus the element's chemical properties, out of line with those of the elements above them in the table. The lack of strong relativistic effects in seaborgium implies that it is going to be hard to predict and understand the way these new elements behave.

To glean information of this sort, chemists have had to refine their techniques of analysis to deal with samples of almost vanishing proportions. The researchers working on seaborgium found their results by conducting chemical reactions on just *seven atoms* in the short time before they decayed. The Berkeley team and others are now hard at work looking at the chemistry of elements 107 and above.

While earlier element-hunters often had to resign themselves to working with microscopic amounts of material, today's pioneers of the Periodic Table thus face the ultimate challenge: to chart the properties of synthetic elements atom by atom.

6

The Chemical Brothers

Why Isotopes are Useful

At first they feared the worst. A body embedded in glacial ice—what else could this be but the gruesome record of a climbing accident? Or perhaps even, to judge from the apparent wound on the back of the dead man's head, something more sinister. Either way, Helmut and Erika Simon, hiking in the Alps along the border of Austria and Italy on 19 September 1991, were deeply disturbed by their discovery.

When informed of the body in the Hauslabjoch pass of the Oetz valley, the Austrian gendarmes assumed that this was another addition to the season's growing list of crevasse accidents. But the corpse was most unusual: its leathery skin was virtually intact, and there was no stench of decomposition. And lying nearby was a strange tool, a kind of primitive axe with a blade of reddish metal.

Maybe this was the long-lost body of an Italian music professor who had reportedly gone missing in the region in 1938? But no: the professor's grave was soon located in a nearby town. It gradually dawned on the forensic scientists investigating the case in Innsbruck that this was a riddle not for them but for archaeologists to solve. The body had been preserved in ice not for decades but for millennia. This man, dubbed Oetzi by the investigators, had died thousands of years ago.

They assumed initially that he must have lived during the Bronze Age, around 2000 BC, since the axe blade looked like bronze. But the true age of Oetzi's body was gauged not from such suppositions but from a scientific measurement. The technique of radiocarbon dating showed that he had died much earlier, around 3300 BC. The axe blade turned out to be made from copper, smelting of which predates the invention of bronze. Copper is soft and was thought to be little used for tools. Oetzi's implement challenged that preconception.

Archaeology was transformed by the invention of radiocarbon dating in the late 1940s. It enables anything made from organic matter—mummified bodies, wooden artefacts, deep-sea sediments—to be dated with generally great accuracy, provided that it is between about 500 and 30,000 years old. Conveniently, this is precisely the period that most archaeologists study—before reliable historical records but after humans began to form societies.

Radiocarbon dating relies on the fact that carbon exists naturally in several isotopic forms. All of them are virtually identical chemically, but they can be distinguished with special methods of analysis. One isotope, carbon-14, provides a kind of elemental clock that reveals the age of carbon-rich materials from living organisms. This technique is one of the most valuable of the many uses that chemists, geologists, medical biologists, and other scientists have found for *isotopes*: the sibling forms that every element displays.

Rounding up the elements

Isotopes answer a puzzle that troubled chemists ever since Dalton proposed his atomic theory. Dalton said that the key property of an atom is not its size or shape but its weight. Each element is characterized by an atomic weight defined relative to that of hydrogen. The fact that these relative atomic weights were usually whole numbers, more or less (carbon's is 12.011, oxygen's 15.999), led Prout to suppose that all elements might be made from hydrogen. The steadily increasing weights gave Mendeleyev, Meyer, and others a kind of index by which to order the elements and reveal their periodic behaviour.

But not all elements conformed so neatly to this picture. Chlorine, for instance, has a relative atomic weight of 35.45, which is close neither to 35 nor to 36. This forced Dumas to conclude that the basic building block of the atom might be smaller than a hydrogen atom. But, with atomic weights like 24.3 and 28.4 (as listed for magnesium and silicon in Mendeleyev's revised 1902 table), how small do you go? Furthermore, Mendeleyev was compelled to place tellurium and iodine out of sequence in ascending atomic weight in order to maintain the periodicities in his original table. And cobalt and nickel seemed to have the same atomic weight!

Francis Aston explained all this in 1919 using his 'mass spectrograph'. Previously, chemists had weighed elements trillions upon trillions of atoms at a time. Aston's machine was able to sort out moving atoms one by one according to their mass, by making them into electrically charged ions and using electric fields to bend their trajectories. He found that atoms of the same element possessed a range of different masses, each one of which was indeed an integer multiple of the mass of the hydrogen atom (that is, essentially the mass of the proton). Sulphur atoms, for example, could have masses of 32, 33, and 34 times that of hydrogen.*

Within two decades of inventing the mass spectrograph, Aston succeeded in identifying 212 of the 281 naturally occurring isotopes of all the elements. He realized that atomic weights measured from bulk samples of an element are averages of the various isotopic forms, which depend on their relative proportions. Thus neon has an atomic weight of 20.2 because it consists of nine parts of the isotope neon-20 mixed with one part of neon-22. For these discoveries, Aston was awarded the chemistry Nobel Prize in 1922.

Isotopes of an element all have the same number of protons in their nuclei (and of electrons orbiting them), but differ in their number of neutrons. Neon-20 has ten protons

---

* As we saw earlier, even these masses are not *exactly* whole numbers. Aston's instrument could measure masses very accurately, and he generally found around a 1 per cent deficit in mass relative to a whole number of hydrogen atoms. This missing mass is converted into the binding energy of the nucleus (see page 131).

(an atomic *number* of 10) and ten neutrons; neon-22 has ten protons and twelve neutrons. The *atomic mass* of an isotope is the total number of protons and neutrons in its nucleus: here 20 and 22 respectively. Chemists denote a particular isotope of an element by writing its atomic mass as a superscript before the elemental symbol: ^{20}Ne, ^{22}Ne.

The chemical behaviour of an element depends on its electrons: how many of them, and how arranged in their shell structure. The configuration of electrons is the same for all isotopes of an element—adding extra neutrons to a nucleus has essentially no effect on the electrons. So isotopes show the same chemical behaviour as one another.

Or do they? There are actually small but sometimes crucial differences in the way isotopes behave. A chemical bond between two atoms is a bit like a spring linking two weights. The vibrations of the spring depend on the masses of the weights: big weights have more inertia, and vibrate more slowly. So the bond vibrations of atoms of different isotopes have slightly different frequencies. Because these vibrations can determine how easy it is to make or break a bond, there are subtle differences in the chemical reactivity of different isotopic forms of an element. Generally these differences are too small to be significant—but not always.

The 'isotope effect' on chemical behaviour is particularly pronounced for hydrogen. This element has three isotopes: 'normal' hydrogen (^{1}H), deuterium (^{2}H, often denoted D)*,

---

* It is not normal to give an isotope a different chemical symbol. But deuterium and tritium are rather special cases.

which occurs naturally in the proportion of about 0.000015 per cent, and tritium (^{3}H or T), which is unstable and decays radioactively. Deuterium has a nucleus containing one proton and one neutron, and so it is twice as heavy as ordinary hydrogen, which contains only a proton. This is why deuterium is called 'heavy hydrogen', and why water containing mostly deuterium instead of 'light' hydrogen (D_2O) is called 'heavy water'.

Doubling the mass of the hydrogen atom has a pronounced effect on its bond vibrations and bond strengths.* The unique properties of water, which make it so vital for life, stem from the way the hydrogen atoms mediate weak attractions between water molecules. These attractions are called hydrogen bonds. In heavy water the hydrogen bonds are slightly stronger, and this changes the properties of the liquid sufficiently to disrupt the lubricating effect water has on biochemical processes. Thus heavy water is a potent poison. The American chemist Gilbert Lewis found in 1931 that tobacco seeds watered with heavy water would not germinate and that mice given small quantities of it showed 'marked signs of intoxication'. Ernest Lawrence, who was

---

* It is tempting to suppose that the bonds formed by deuterium are stronger than those formed by hydrogen simply because deuterium is more sluggish and thus slower to react. But the big isotope effect in this case stems from a more subtle and complex effect, rooted in quantum mechanics. In essence, the lighter hydrogen atom is able to 'tunnel' its way out of a bond, reflecting its quantum-mechanical wavelike nature, more easily than deuterium. Quantum effects like this are seldom significant for other elements; they occur for hydrogen because it is so small and light.

eager to use the rare and valuable deuterium nuclei in his cyclotron experiments, was disgusted that Lewis saw fit to feed it to mice.

The age of carbon

Isotopes *can* differ significantly in one respect: the stability of their nuclei. The nucleus of a carbon atom, for instance, will happily accommodate six or seven neutrons alongside its six protons; but fewer or more neutrons make the nuclei unstable and liable to decay radioactively. Nuclear reactions, like those conducted in Lawrence's particle accelerators, can convert stable nuclei into unstable ones. Even 'benign' elements like carbon and nitrogen can be converted into radioactive and potentially hazardous forms in this way.

Indeed, this happens every moment in the Earth's atmosphere. The upper atmosphere is bombarded with cosmic rays: fast-moving subatomic particles produced by extremely energetic astrophysical processes such as nuclear fusion in the sun. When cosmic rays hit molecules in the atmosphere, they induce nuclear reactions that spit out neutrons. Some of these neutrons react with nitrogen atoms in air, converting them into a radioactive isotope of carbon: carbon-14 or 'radiocarbon', with eight neutrons in each nucleus. This carbon reacts with oxygen to form carbon dioxide. About one in every million million carbon atoms in atmospheric carbon dioxide is ^{14}C.

Carbon-14 decays by emitting a beta particle, transforming it back into the most stable isotope of nitrogen. But it is in no hurry to do so: the half-life of ^{14}C is around 5,730 years. This time scale makes radiocarbon the ideal archaeologist's tool.

Carbon is constantly taken in by living organisms. Plants pluck it from the air and fix it in their tissues by photosynthesis. Animals consume the carbon compounds of plants and other animals. The flux of carbon through living bodies means that they maintain a more or less constant, minuscule level of radiocarbon.

When an organism dies, it stops acquiring new carbon, and the amount of radiocarbon it contains begins to decline through radioactive decay. Wood from a tree that died (when felled for timber, say) 5,730 years ago has only half as much radiocarbon as that from a similar tree felled recently. Wood that is 11,460 years old (assuming it is somehow preserved) has only a quarter as much. So by measuring the ^{14}C content of ancient wooden artefacts we can deduce how old they are. The same applies to bones, to cloth and paper and animal fat used to bind pigments in cave paintings. The measurement is done in a mass spectrometer, an instrument like Aston's spectrograph, which separates the different isotopes of carbon.

The American chemist Willard Libby realized in 1947 that ^{14}C could be used for archaeological dating. Libby studied radiochemistry at Berkeley in the 1930s and subsequently worked on the Manhattan Project. After the war he joined the Institute of Nuclear Studies in Chicago, where Fermi made the first nuclear reactor. Libby and his co-workers tested their dating technique on wood and charcoal found in Egyptian graves, whose age was already well known to archaeologists from historical analysis, and on very old redwood trees that could be independently dated by counting tree rings. Libby's technique was used to date the end of the last Ice Age and the creation of human settlements in regions ranging from North America to Iraq. The invention of radiocarbon dating earned Libby the Nobel Prize in chemistry in 1960.

Radiocarbon dating has often become the final arbiter in contentious archaeological and historical debates. Few have been more contended than the study of the Shroud of Turin, reputedly the cloth that wrapped the body of the crucified Christ (Fig. 14). The shroud is imprinted with the image of a naked man who bears the marks of whipping and crucifixion. It was investigated scientifically in the 1970s, but radiocarbon dating was not used because the methods available at that time required an unacceptably large amount of material.

In 1988 teams from Arizona, Oxford, and Zürich collaborated to establish the shroud's true age through sensitive radiocarbon measurements made at the expense of three small patches of cloth weighing just 50 milligrams each. The results indicated that, with high probability, the cloth was made some time between AD 1260 and 1390. The shroud was, it seemed, a medieval forgery.

It is no surprise that these claims have been disputed, for the shroud carries immense iconic significance. One complaint is that the cloth may have become contaminated over

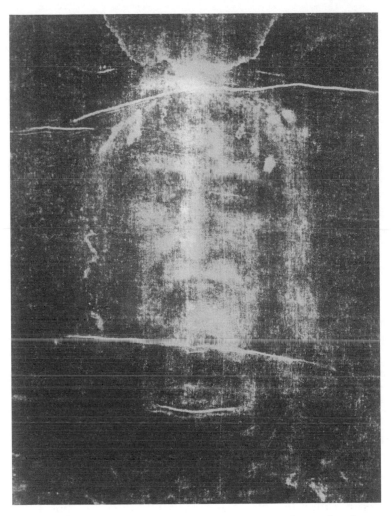

14 Radiocarbon dating of the Shroud of Turin indicates that it was manufactured in the thirteenth or fourteenth century

the years by fungal growth and organic materials deposited by bacteria, as well as by smoke from a well-documented fire in 1532 in the French chapel at Chambery where it was then housed. Certainly, such complications have occasionally invalidated radiocarbon dating of other items in the past. And still no one knows quite how the image was imprinted on the cloth, nor how it came to be so historically and anatomically accurate (medieval artists conventionally showed Christ's stigmata on the palms of the hands, whereas nails were driven through the wrists in crucifixion). This is one mystery that radiocarbon will not lay easily to rest.

Dating the universe

If ^{14}C had a half-life of two minutes or a million years, it would be of no use to archaeologists. In the first case it would disappear almost as soon as an organism died; in the second it would barely change over the hundred- to thousand-year time scales that are relevant to human history. To look further back into the past, scientists need radioactive isotopes that decay more slowly.

Many such exist in rocks and minerals, and they enable geologists to reconstruct the history of our planet long before the first humans emerged. Uranium isotopes are some of the most useful geochemical clocks. Uranium-238 decays with a half-life of about 4.5 billion years, almost the same as the age of the Earth. A sequence of decay steps converts ^{238}U to thorium-230.

This is exploited in the technique of uranium-thorium dating, which involves measuring the amount of thorium-230 that has accumulated in a substance by decay of uranium. If the object contained no thorium at all when it was formed, the ratio of remaining ^{238}U to accumulated ^{230}Th is a measure of the age. The object being dated must not have had access to sources of 'fresh' uranium that could reset the clock. This is true, for example, of coral left stranded on 'fossil beaches' when sea levels recede, or of stalagmites and stalactites in caves. Wood and fossil bone have also been dated this way. Because thorium-230 itself decays with a half-life of just 75,380 years, the U-Th method cannot provide accurate dates further back than 500,000 years or so.

The decay of ^{230}Th leads to radioisotopes of other elements, ultimately concluding with the stable isotope lead-206. Happily, some of the oldest rocks on Earth, called zircons, contain no lead when they are formed. This means that the amount of lead they accumulate over time from uranium decay reflects their age. Until the rocks crystallized, uranium atoms could move freely through the molten magma from which they formed, and decayed uranium could be replenished. Solidification of a zircon does for uranium what an organism's death does for radiocarbon: it stops the influx of fresh radioactive material, and the decay clock starts ticking. Because of ^{238}U's long half-life, zircons can be dated back to the Earth's earliest days.

Our planet was probably transformed to a ball of magma 4.45 billion years ago by an impact with a small planet-like body; the resulting debris formed the Moon. Yet uranium-lead dating shows us how quickly this 'magma ocean' must have cooled, since it reveals that the oldest zircons, found in Western Australia, crystallized about 4.4 billion years ago. What is more, these ancient zircons show signs of having been formed in contact with water, implying that even in that distant era the world had oceans.

A small proportion of natural uranium consists of the isotope ^{235}U. This decays not to lead-206 but to lead-207. By measuring the amounts of all these isotopes of uranium and lead in rocks, geologists can date all manner of minerals, and can even reconstruct the history of our planet's formation. Some meteorites are thought to be left-over remnants of the rocky material that aggregated to produce the Earth, and they show us the mixture of elements this material contained. If they contain no uranium, then all the lead in these meteorites must be 'primordial'—it must have been there at the outset, rather than being generated by decay of uranium. By comparing isotope ratios in old lead ores to those in such meteorites, scientists can figure out how old the meteorites are. As they are coeval with the Earth, such measurements put a date on the formation of our planet. This happened about 4.54 billion years ago.

The American scientist B. B. Boltwood appreciated as early as 1907 that radioactive decay can tell us about the age of the Earth. The best estimate until that time was around 98 million years, which Lord Kelvin deduced in the 1860s by considering how long it would take for the hot core to cool down.* Boltwood calculated that the planet could be as much as two billion years old. The current estimate of more than twice this value is supported by a host of other 'radiometric' methods which look at the relative abundances of 'parent' and 'daughter' isotopes in radioactive decay chains.

Many other pairs of isotopes linked by decay processes with long half-lives are used for geological dating of rocks, including samarium-147/neodymium-143, rubidium-87/strontium-87, and potassium-40/argon-40. Each works best for a particular rock type and time scale. Uranium-238 decay has even been used to find the ages of distant stars. In 2001, a powerful telescope at the European Southern Observatory in Chile was used to deduce the abundance of ^{238}U in an old star called CS 31082–001 in our galaxy by measuring the light emitted from uranium in the spectrum of the starlight (see page 91). This study revealed that the star is 12.5 billion years old. The ages of old stars give us a minimum estimate of how long ago the Big Bang happened, since the universe itself must be older than the stars it contains.

---

* Ernest Rutherford, however, used the alpha decay of uranium, which produces helium, to estimate the ages of several uranium ores in 1906. By measuring the ratio of helium to uranium and the current rate of helium production (that is, the current decay rate of uranium), he deduced that the minerals were at least 440 million years old. All this was, of course, before anyone knew anything about isotopes.

History on ice

Radioactive isotopes have thus enabled scientists to reconstruct the history of the Earth and its environs over billions of years. But *stable* isotopes are also an indispensable part of the geoscientist's inventory. In particular, their measurement in the geological record has revolutionized our perception of how the planet's climate system works and how it has changed over time.

This is more than a matter of academic interest. Faced with the likelihood that human activities such as the burning of fossil fuels have altered the world's climate over the past century, we need to know more about the factors that control climate in order to predict what the future might hold. The study of stable-isotope records of the past has shown that the climate system is vastly more complex than anyone dreamed several decades ago, and that it has a capacity for changing its behaviour rapidly and in ways that are hard to anticipate.

Geologists in the nineteenth century deduced that the Earth has experienced several ice ages, during which the ice sheets that cover the poles today reached much farther afield. In 1930 the Serbian mathematician Milutin Milankovitch showed how changes in the shape of the Earth's orbit around the Sun could trigger an ice age by altering the seasonal distribution of sunlight at the planet's surface. There are three cyclic variations in the orbit, with periods of 23,000, 41,000, and 100,000 years. The interplay of these 'Milankovitch cycles' produces a complex but predictable and slow variation in climate over hundreds of thousands of years.

To test Milankovitch's theory, it was not enough to know the dates of a few past ice ages. The theory predicted that the climate system has an unsteady pulse, with ice ages of varying severity, in which the three major rhythms should be discernible. To take this pulse, scientists needed a way to reconstruct a continuous record of how global average temperatures and ice volumes have altered over the past one million years or so.

In the 1970s geochemists realized that such a record might be found in the sediments deposited at the bottom of the oceans. These sediments are formed from the matter that settles out of the ocean water, which is mostly the debris of dead marine organisms. This consists largely of the insoluble mineral shells of microscopic organisms called foraminifera. The shells are made from calcium carbonate, a compound of calcium, carbon, and oxygen. The oxygen comes from the water in which the foraminifera live.

Oxygen has two stable isotopes: ^{16}O and ^{18}O. When seawater evaporates, water molecules containing the lighter isotope escape slightly more easily, just as a sparrow takes flight more easily than an albatross. So evaporation makes the sea richer in ^{18}O. The water vapour soon falls back to earth as rain or snow. Rivers return rainwater to the sea; but in the polar regions snow accumulates as ice, so the water gets locked away for long periods of time. If the ice sheets grow during an ice age, more water vapour is transformed to ice, and seawater gets ever richer in ^{18}O. So the $^{16}O/^{18}O$ ratio of seawater reflects the extent of global ice coverage.

The carbonate shells of foraminifera preserve this isotope

ratio when they are incorporated into sediments, and mass spectrometry can be used to measure the ratio. So the *oxygen isotope record* of deep-sea sediments tells us about past changes in the extent of the ice sheets.* In an international initiative in the early 1970s called the Climate Long-range Investigation, Mapping and Prediction (CLIMAP) project, columns of sediment drilled out of the seabed were analysed to produce a record of climate change over the past 700,000 years. The waxing and waning of the ice sheets, revealed by the changing oxygen isotope ratios in the sediment cores, showed precisely the three dominant rhythms predicted by Milankovitch. The 100,000-year cycle is particularly prominent (Fig. 15a).

The oxygen isotopes in the water molecules locked up in polar ice sheets tell another version of the planet's climate story. The ice that covers Antarctica is a mile and a half deep at its thickest points, and the snow that became transformed into the deepest ice fell at least 250,000 years ago. So the Antarctic ice sheets, like deep-sea sediments, encode thousands of years of climate history in their isotope compositions.

* It was initially thought that the oxygen isotope ratio of sediments was a measure of the *temperature* of the seawater, since this affects how the two oxygen isotopes are apportioned when oxygen is transferred from water molecules to carbonate as the foram shells develop. But studies in the 1960s and 1970s showed that the oxygen isotope ratios in ocean sediments are controlled mostly by changes in the global volume of ice sheets. The scientists also worked out how to use these oxygen isotope records to deduce changes in the temperature of water at the sea surface, and found that in the tropics this did not differ much during and after the last ice age.

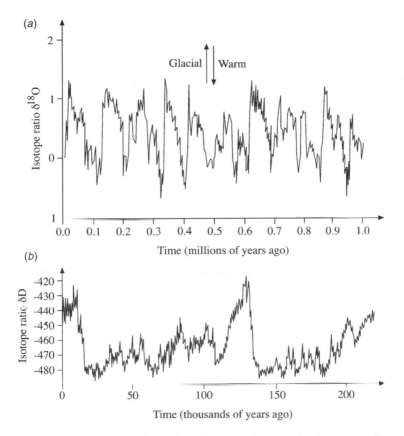

15 *a.* Measurement of the ratios of oxygen isotopes in deep-sea sediments tells us how global climate has fluctuated in the past. The changes in the isotope ratio, defined as a quantity called $\delta^{18}O$, reflect the changes in volume of the world's major ice sheets. When the ice volume is high—during ice ages—$\delta^{18}O$ is large. The climate record shown here for the past million years, deduced from a sediment core in the eastern tropical Pacific Ocean, has a jittery pulse in which, at least for the past 700,000 years, an oscillation that rises and falls every 100,000 years is evident. *b.* The ratios of both oxygen and hydrogen isotopes (H and D, quantified by the parameter δD) in polar ice sheets provide another source of climate records. In this hydrogen isotope record from the Vostok ice core in Antarctica, δD reflects how air temperatures over the ice sheet have changed in the past several thousand years: a high δD indicates a relatively warm period

The oxygen isotope ratio of ice cores is mostly controlled by a different influence, however: the temperature of the clouds from which the snow fell. When water vapour condenses to water or ice, isotope sifting occurs just as it does during evaporation—but in reverse: the lighter isotope stays behind. The last precipitation to leave a cloud—the snow that falls over the poles—is therefore enriched in ^{16}O. The amount of enrichment turns out to depend on how cold it is over the ice sheet. So ice-core isotope records show us how atmospheric temperatures have changed over time.

Ice cores have been drilled at several places in the Antarctic, including the research outposts at Vostok and Byrd Station. What they tell us is largely consistent with the climate records obtained from the Greenland ice sheets on the other side of the world, as well as with those from marine sediment cores. One can double-check these ice-core records because the ratio of ^{1}H to deuterium in the water molecules of the ice also acts as an atmospheric thermometer (Fig. 15b).

As recorders of past climate change, ice sheets have an advantage over marine sediments. Bottom-dwelling creatures in the sea stir up the top inch or so of sediment, blurring the isotope record. Every layer in the sediment has been disturbed this way as it was formed. The snow on the ice sheets, meanwhile, is undisturbed as it gets compacted into ice. This means that the ice-core records show more of the fine detail of temperature changes. Ice-core climate records reveal that temperature shifts can be amazingly rapid. In some cases the climate of the North Atlantic region seems to have switched from ice-age to warm (interglacial) conditions in the space of just a few decades. This is much faster than can be accounted for by Milankovitch cycles, and is thought to reveal a switch-like instability of the Earth's climate system, probably due to changes in the way water circulates in the oceans.

Polar ice contains tiny bubbles of trapped ancient air, within which scientists can measure the amounts of minor ('trace') gases such as carbon dioxide and methane. These are greenhouse gases, which warm the planet by absorbing heat radiated from the Earth's surface. The ice cores show that levels of greenhouse gases in the atmosphere, controlled in the past by natural processes such as plant growth on land and in the sea, have risen and fallen in near-perfect syn-chrony with temperature changes. This provides strong evidence that the greenhouse effect regulates the Earth's climate, and helps us to anticipate the magnitude of the changes we might expect by adding further greenhouse gases to the atmosphere.

Healing with radiation

While he was investigating radioactive isotopes with Ernest Rutherford in 1913, George de Hevesy had an idea. Nuclear scientists were commonly forced to work with only tiny quan-tities of radioactive material, which would be very difficult to 'see' using standard techniques of chemical analysis. But every single atom of a radioisotope advertised its presence when it decayed, since the radiation could be detected with a Geiger counter. So, if a radioisotope of an element could be concentrated by separating it from stable isotopes of the same element, it could be used in tiny quantities as a kind of marker that tracks the movements of a substance. It would behave chemically just like the 'normal' element but would betray its presence by emitting radiation.

De Hevesy realized that this radioactive-marker technique could be particularly valuable for biological studies: for following the progress of chemicals through the human body. Alpha and beta particles are absorbed by organic tissues, but gamma radiation can pass through several feet of concrete and so has no trouble escaping the body. Once the Joliot-Curies had shown that radioisotopes of any element could be made artificially, it became possible to find all sorts of gamma emitters suitable as 'tracers' for studying biochemical processes.

Phosphorus-32, for example, produced by irradiating sulphur or natural phosphorus (^{31}P) with high-energy particles, has a half-life of 14.8 days and can be rapidly taken up (in the form of phosphate) by body tissues such as muscles, the liver, bones, and teeth. De Hevesy found that different phosphorus compounds would be incorporated in a tissue-specific manner: certain compounds were concentrated in the liver, for example. One can use stable isotopes as biological tracers too, since they are detectable atom by atom using mass spectrometry. De Hevesy observed that it takes deuterium twenty-six minutes to pass from ingested heavy water into urine.

De Hevesy's work launched the use of isotopes in biology and medicine, and it won him the 1943 Nobel Prize in chemistry. Our normal instincts are to put as much distance as possible between ourselves and radioactive substances, for indeed radiation can be lethal.* But the poison is in the dose, as Paracelsus was fond of saying. Radioactive isotopes can be used as tracers at concentrations too low to pose any health hazard.

An isotope of the rare element technetium, denoted ^{99m}Tc, is widely used to form images of the heart, brain, lungs, spleen, and other organs. Here the 'm' indicates that the isotope, formed by decay of a radioactive molybdenum isotope created by bombardment with neutrons, is 'metastable', meaning only transiently stable. It decays to 'normal' ^{99}Tc by emitting two gamma rays, with a half-life of six hours. This is a nuclear process that does not change either the atomic number or the atomic mass of the nucleus—it just sheds some excess energy.

As a compound of ^{99m}Tc spreads through the body, the gamma radiation produces an image of where the radioisotope has travelled. Because the two gamma rays are emitted simultaneously and in different directions, their paths can be traced back to locate the emitting atom precisely at the point of crossing. This enables three-dimensional images of organs to be constructed (Fig. 16). Scientists are devising

---

* It was not always perceived this way. In the early twentieth century, radium, which killed Marie Curie, was sold as a cure-all, leading *Nature* to warn that 'there is a danger that the claims which have been advanced for radium as a curative agent may lead to frauds on the credulous section of the public'.

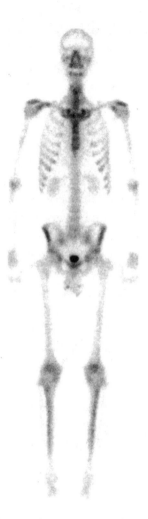

16 An image of the human body recorded from the radioactive decay
of metastable technetium-99 in the bloodstream

new technetium compounds that remain localized in specific organs. Eventually, the technetium is simply excreted in urine.

Making metastable technetium-99 is an expensive business. A cheaper, common alternative tracer is iodine-131, which emits a gamma ray when it decays. But the iodine isotope also releases beta particles that can damage tissues, making it less attractive as an imaging agent.

Another form of three-dimensional imaging of internal organs, called positron emission tomography (PET) scanning, exploits a less common form of beta decay. Most beta decays involve the emission of electrons from the nucleus as a neutron decays into an electron and a proton. But the reverse can happen too: a proton can decay into a neutron (see page 132). The positive charge is borne away by a positron, which will soon collide with an electron. Their mutual annihilation produces a gamma ray.

Positron-producing beta decay happens to neutron-poor nuclei. Two such are the isotopes carbon-11 and fluorine-18, which are short-lived isotopes produced in nuclear reactors. In PET scanning, compounds of these isotopes are ingested and the gamma rays produced by positron-electron annihilation in the body (which happens very close to the point of positron emission) are used to construct three-dimensional images as a series of two-dimensional slices. PET scanning is particularly useful for brain imaging.

The damaging effects that radioactivity can have on tissues are not all bad. To treat cancers, we *want* to kill cells—albeit the unhealthy, frantically replicating tumour cells, not healthy cells. If radioisotopes can be localized in tumours, they do their destructive business to good effect. Cobalt-60, made by neutron bombardment of stable cobalt-59, is a radioisotope with a half-life of 5.3 years that is used to treat cancer.

The cobalt nucleus decays to nickel-60 by emitting a beta particle and two gamma rays. The gamma rays do most of the damage; even though they pass through human tissues, occasionally they will knock an electron from an atom in a cell and set in train a series of biochemical 'free radical' reactions that can trigger the death of cells. In cancer treatment the aim is to ensure that cobalt-60 gets selectively to the tumour. Unfortunately this targeting remains imperfect and some healthy tissue is damaged too. So radiotherapy is a drastic measure for combatting cancer. The dream is to find compounds of this and other radioisotopes that pass straight through the body but gather in cancer cells, providing a 'magic bullet' to knock out only the bad guys.

Gamma rays from cobalt-60 are also used to sterilize food, since they kill bacteria. The gamma rays are incapable of inducing radioactivity within the food, so the method is potentially 'clean'. The rays do, however, produce some free radicals, which are potentially harmful substances. But the concentrations of these are very small, and they may well do less harm than the preservatives otherwise used to protect food from bacterial decay. All the same, radioactivity has an understandably bad image and many shoppers continue to be wary of irradiated food. Of course, the ideal alternative is simply to eat it fresh.

Isotopes are thus a kind of free give-away bonus to the Periodic Table. In a sense they expand our choice of elements by giving us extra versions that do unique and useful things. We do well to remember that each entry in the table represents not a sole member of the element family but a kind of averaged image of a small group of chemical brothers and sisters, each with their own talents.

7

For All Practical Purposes

Technologies of the Elements

Of the elements that have shaped the fates of civilizations, arguably none has been more instrumental than the martial metal, that most stable of elements: iron. The Hittites of Asia Minor in the thirteenth century BC seem to have been the first culture to smelt and forge iron systematically, and this gave their armies an edge, quite literally, over their rivals. The militaristic Assyrians mastered the art around the ninth century BC, and no opponent could resist their brutal iron fist for several centuries.

Rome dug and traded far and wide for the iron that equipped its legions with keen-bladed swords and gleaming armour. This shining metal was not raw iron but hard steel, which bent the softer wrought-iron blades of the Gauls. Steel of a sort was made by the Hittite smiths, by hammering and heating the iron in contact with charcoal: a process called cementation. Tempering—plunging the hot metal into cold water—made steel harder still. The finest steel in the Roman Empire was so-called Seric iron, forged in southern India and imported through Abyssinia.

The use of charcoal in steel making thus has a long history. Yet it was not until the eighteenth century that steel's key additive—carbon—was identified. Since charcoal was traditionally used to smelt iron from its ore, some carbon was always incorporated into the metallic product by chance. But the proportion of carbon determines the hardness of the product, a fact noted by the Swedish metallurgist Tobern Bergmann in 1774. Controlling the carbon content in steel was an erratic process until the Englishman Henry Bessemer invented his steel-making process in the 1850s. In the late nineteenth century steel transformed construction engineering, and at the beginning of the twenty-first century the world market for steel was estimated at about $500 billion.

Steel is no longer simply a matter of spicing iron with the right amount of carbon. Stainless steel contains at least 10 per cent chromium, and high-performance engineering steels might incorporate purposeful additions of nitrogen, phosphorus, sulphur, silicon, nickel, manganese, vanadium, aluminium, titanium, niobium, molybdenum, and other elements besides. The properties of the metal are fine-tuned by an intricate mixture of elemental ingredients.

The Iron Age, then, is something of a misnomer. Not only was iron used long before the Iron Age dawned,* but it was really the invention of steel that turned nations into

* Some historians date the Iron Age to around 1200 BC, when the Hittite empire was destroyed and its smiths were dispersed, spreading the knowledge of ironworking. But man-made iron artefacts existed before 2500 BC. The Iron Age, along with the earlier Bronze and Stone Ages, is an invention of nineteenth-century archaeologists and of questionable value today.

conquerors. Myth and symbol, however, attach less readily to an elemental mélange: it is an *iron* horse that steamed its way across the American plain, the iron fist that represents a display of might. Oliver Cromwell's steel-clad Ironsides crushed the Royalist troops of Charles I, the Iron Cross honoured German military valour, the Iron Curtain marked the boundary of cold war national alliances. After all, elemental iron's glittering grey strength distinguishes it from copper's rosy malleability or soft yellow gold. Iron can be improved, but it is the characteristic properties of the element itself that mark it out for battle and conjoin it with Mars, god of war.

Many other elements find applications that are uniquely determined by their fundamental nature. In this final chapter I shall consider some of them. It is a fairly random selection, for just about every nook of the Periodic Table has been explored for what it can offer to our advantage. I hope to give, by way of conclusion, a flavour of the variety that exists among the elements and a sense of why this provides countless opportunities for making useful things from the riches on Mendeleyev's table.

Chips with everything

If one single element divides the modern world from that before the Second World War, it is the unassuming grey solid called silicon. This element is everywhere, and always has been. Silicon is the second most abundant element in the

Earth's crust, since most common rocks have crystalline frameworks made from silicon and oxygen: they are silicates. Quartz and sand are composed of silicon and oxygen alone: silicon dioxide, or silica.

These natural compounds of silicon are the raw material for the oldest technology: stone tools more than two million years old have been found in Africa. Some time around 2500 BC, Mesopotamian artisans found that sand and soda could be melted in a furnace to produce a hard, greenish translucent substance: glass. They coloured it with metal-containing minerals and used it to make gorgeous vessels and ornaments. Glass-making was improved in the Middle Ages when craftsmen discovered how to remove the greenish tint (due to iron impurities). To awed churchgoers, the multicoloured windows telling the stories of the Gospels in glowing light must have been as captivating as a modern movie. And the perfecting of grinding methods for making lenses opened up the heavens to Galileo and his contemporaries, bringing a concrete reality to the previously immaculate celestial realm. Glass, it can be argued, changed the view of our place in the universe.

For a long time, silica was considered to be an element—Lavoisier lists it as such—for it is not easy to persuade silicon and oxygen to part company. Humphry Davy suspected that silica was not elemental, but silicon itself was not isolated until 1824, when Jons Jacob Berzelius prepared it in a form called amorphous silicon. This is a solid in which the atoms are not regularly arrayed as they are in a crystal, but are more jumbled. Glass is also amorphous, its silicon and oxygen atoms in mild disarray. Crystalline silicon was not made until 1854, by the French chemist Henri Deville.

But it took us a very long time to figure out what this pure silicon is good for. It occupies that curious no man's land in the Periodic Table where metals (to the left) give way to non-metals (to the right). Silicon is not a metal, but it does conduct electricity—albeit poorly. It is a semiconductor.

Technically this means rather more than 'bad conductor'. Metals conduct electricity because some of their electrons come free of their parent atoms and are at liberty to roam through the material. Their motion corresponds to an electrical current. A semiconductor also has wandering electrons, but only a few. They are not intrinsically free, but can be shaken loose from their atoms by mild heat: some are liberated at room temperature. So a semiconductor becomes a better conductor the hotter it is. Metals, in contrast, become poorer conductors when hot, because they gain no more mobile electrons from a rise in temperature and the dominant effect is simply that hot, vibrating atoms obstruct the movement of the free electrons.

Since electronics is all about moving electrical currents around, it may seem strange that a semiconductor rather than a metal is used to make the electrical components on silicon chips. But silicon's paucity of 'conduction electrons' is the whole point here. It means that the conductivity can be delicately fine-tuned by sprinkling the crystal lattice with atoms of other elements, which increase or decrease the number of mobile electrons. In a metal, awash with mobile electrons, this would be like trying to adjust the water level of a raging river by emptying into it a few brimming thimbles.

Arsenic atoms have one more electron than silicon atoms in their outer shell. So 'doping' silicon with arsenic injects a precious few extra electrons: one for every arsenic atom. Likewise, boron has one fewer electron than silicon, so boron doping reduces the number of conduction electrons. This does not actually make boron-doped silicon a poorer conductor, since an electron deficiency in a silicon crystal lattice introduces a kind of hole in the 'electron sea', like a gap in a crowd. This hole can move around just as a free electron can, but it acts as though it has the opposite (positive) charge. So arsenic doping of silicon adds mobile electrons—negatively charged agents of the electrical current—and it is called n-type doping. Boron-doped silicon contains positive charge carriers, and is called p-type.

Microelectronic devices on silicon chips are typically made from layers of n-type and p-type silicon. Films of silica act like the plastic sheath on copper cable, since silica is insulating. A layer of p-type silicon back to back with a layer of n-type, called a p–n junction, allows a current moving across the junction to flow in one direction but not the reverse. This one-way behaviour is the fundamental characteristic of a device called a diode. Early diodes in electronics were made from metal plates sealed inside evacuated glass tubes, which could be seen glowing in the innards of old radio sets. Diodes made from doped silicon can be much smaller and more robust: since they are made from solid materials, they are components of 'solid-state' electronics.

The workhorse of silicon-based electronics is the transistor. This is a slightly more complex sandwich of p-type and n-type layers, creating a device through which the electrical current can be controlled by an applied voltage. This gives a transistor the ability to act as a switch, turning signals on and off, and also as an amplifier that creates a strong signal from a weak one. Transistors can be built into circuits capable of performing 'logic' operations, such as the fundamental mathematical processes of addition and subtraction. Logic circuits are wired together on silicon chips to make microprocessors and computers.

The first solid-state transistor was made not from silicon but from the element below it in the Periodic Table: germanium. This substance is also a semiconductor, and can be doped in the same way. William Shockley, Walter Brattain, and John Bardeen devised the germanium transistor at Bell Telephone Laboratories in New Jersey in 1947. It was a crude and clunky device (Fig. 17a)—bigger than a single one of today's silicon chips, which can house millions of miniaturized transistors, diodes, and other components (Fig. 17b). The three inventors shared the Nobel Prize in physics in 1956.

Silicon for chip manufacture must be highly pure and free of defects in the crystalline packing of atoms. It is made by a technique developed in the 1940s called Czochralski growth, in which silicon extracted from quartz and purified is melted and drawn out slowly into rods. The rods are sawn into slices, providing the silicon wafers on which a chip's circuitry is constructed. A cheaper way of making crystalline silicon,

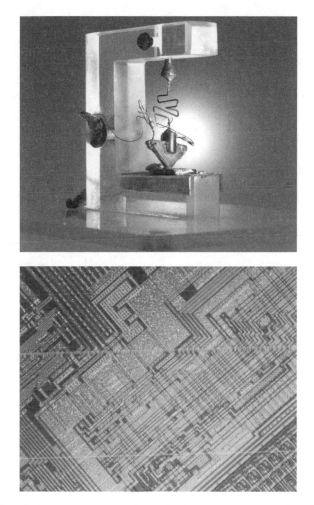

(a)

(b)

17 The first prototype transistor (a 'point-contact semiconductor amplifier'), built by Bardeen and Brattain at Bell Laboratories in 1947 (a), is a far cry from today's silicon chips, packed with miniaturized semiconductor components (b)

called the Wacker process, was invented in the 1970s. It casts molten silicon in moulds, just as metal components are cast. The resulting crystals are riddled with flaws: they are really a patchwork of tiny crystallites welded together with their atomic lattices tilted at different angles. This 'polycrystalline' silicon is not much use for electronics, since the flaws degrade the conductivity; but it is used, for example, to make silicon solar cells, which are the most common commercial photovoltaic devices. In these devices, sunlight absorbed by thin films of silicon kicks electrons free from their parent atoms, producing pairs of electrons and holes. These are collected at two electrodes, creating a flow of current.

Among the useful compounds of silicon are silicon carbide (carborundum) and silicon nitride, which are hard, tough materials used for making cutting tools, abrasives, and engineering components resistant to heat. In stark contrast, silicon and oxygen can be fashioned into soft materials called silicones that contain long chains (polymers) in which the two types of atom alternate. This ability to form chainlike molecules is rare. Carbon is the chain-former *par excellence*, which is why it is the central ingredient of complex organic molecules. The chains of silicone compounds are very stable, however, which makes them robust and versatile engineering materials.

Some silicone polymers are slippery oils, which are used as lubricants, paint binders, and fluids for cosmetics and hair conditioners. The longer the chains, the more viscous the oil. By linking the chains to one another at various points to form a network, silicones can be solidified into soft rubbers and resins. Silicone rubber is the ideal sealant for kitchens and bathrooms, as it is non-toxic and water-repellent. Its non-flammability recommends it for fire-fighting suits, and it gained a little glamour in 1969 when Neal Armstrong took his small step for a man wearing silicone boots.

On the other hand, silicone's reputation nosedived when Dow Corning, the major manufacturer, was forced to hand out billions of dollars of compensation in response to lawsuits claiming that leakage of silicone breast implants had damaged the health of many women. These implants contain silicone oil within a sac of rubbery silicone. The charge was that silicone had led to autoimmune diseases in implantees. There is still no clear evidence that the compound is harmful in any way, but nevertheless in 1992 a moratorium was imposed in the USA on its use for implants.

A new kind of silver

When palladium was first discovered, no one seemed to want it. Its discoverer, William Hyde Wollaston, offered it for sale in a London shop as 'new silver', at six times the price of gold. Hoping to profit from his discovery, at first he chose not to disclose to the scientific community how he obtained the metal. But there were few takers for 'new silver', and Wollaston eventually took back most of his stock of palladium and donated it to the Royal Society, where he announced the preparation and properties of the new metal in 1805.

It did truly look like silver. Furthermore, it was malleable enough to be made into jewellery, and resisted the corrosion that gradually turned real silver black. In this respect palladium closely resembles platinum, which sits below it in the Periodic Table. It is in fact one of the so-called platinum-group metals, all of which were found lurking in natural platinum around the turn of the nineteenth century by Wollaston and his colleague Smithson Tennant.*

In the course of investigating the production of platinum from its ores, Wollaston and Tennant found four new elements in 1803. Tennant isolated osmium and iridium; Wollaston found rhodium and palladium. As was the contemporary habit, Wollaston named the latter after a newly discovered celestial body. Uranium gained its name this way after William Herschel's discovery of the planet Uranus, and palladium honoured the asteroid Pallas, found in 1802.

Only recently did palladium find its niche. All of the platinum-group metals are good catalysts: they speed up the rate of certain chemical reactions. Simple gases such as oxygen and carbon monoxide become stuck to the surfaces of these metals, whereupon they fall apart into their constituent atoms. The atoms then wander around on the surface until they encounter others, combining in new configurations.

Platinum, palladium, and rhodium all catalyse reactions that transform some of the noxious gases in the exhausts of cars into less harmful compounds. Carbon monoxide, a potent poison, may be transformed in this way to carbon

---

* Tennant was the first to show, in 1797, that graphite and diamond are composed of the same pure element—carbon.

dioxide, and unburnt hydrocarbons from the fuel get burnt up on the metal surfaces. Nitric oxide, one of the main contributors to urban smog, will react with carbon monoxide to form carbon dioxide and nitrogen gas. These processes are conducted in catalytic converters.

Inserted into the exhaust system of vehicles, catalytic converters can reduce emissions of carbon monoxide and hydrocarbons by up to 90 per cent. The first catalytic converters used mainly platinum, but now palladium is the predominant catalytic metal. The metals are dispersed as tiny particles on a supporting framework of porous aluminium oxide (alumina) (Fig. 18).

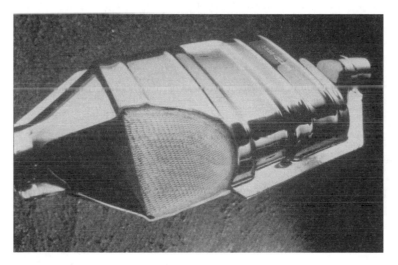

18 Catalytic converters use palladium and related metals to rid motor exhaust of its noxious gases

Sixty per cent of the palladium manufactured worldwide—mainly as a by-product of nickel, zinc, and copper refining—is now used in catalytic converters. Much of the rest is used in electronic components, but a little is used for jewellery, showing that we have after all acquired a taste for Wollaston's 'untarnishable silver'.

In 1989 palladium prices temporarily soared. Two chemists at the University of Utah, Martin Fleischmann and Stanley Pons, claimed that it was the key to a cheap method for transforming hydrogen to helium by nuclear fusion, producing a new, safe, and clean source of energy. Stock-market investors figured that this 'cold fusion' process was about to make palladium hot stuff. Political strategists, meanwhile, worried that the countries with the major mineral resources of palladium—South Africa and the Soviet Union—might find themselves in an unexpectedly powerful position.

What no one seemed to realize until later was that we had been here before. In the 1920s two German scientists, Fritz Pareth and Kurt Peters, proposed that hydrogen might be converted into helium inside palladium metal. Their aim was not to create an energy source; it was the helium they were after. Since the demise of the hydrogen-filled Hindenburg, helium was in big demand as the buoyant gas for airships.

Pareth and Peters knew that palladium acts as a kind of hydrogen sponge, absorbing huge quantities of the gas. At room temperature, palladium will accommodate more than 900 times its own volume of hydrogen. The hydrogen molecules fall apart into two separate atoms on the surface of the metal, and the tiny hydrogen atoms can diffuse into the spaces between metal atoms. The metal expands by up to 10 per cent as it soaks up hydrogen, setting up huge internal pressures. Might these be big enough to squeeze two hydrogen atoms together to make helium? When the researchers tested the idea using a palladium wire, they found minute traces of helium.

The news reached John Tanberg in Sweden, who was later to become scientific director of the Electrolux company. He suspected that the apparent production of helium might be speeded up by using electrolysis. This involves inserting two oppositely charged electrodes into a liquid that contains ions, such as a solution of a salt. The positive ions are attracted to the negative electrode and vice versa. An acidic solution contains positively charged hydrogen ions, and Tanberg reasoned that a negative charge applied to a palladium plate might cram these ions at high density into the metal. He too found helium when he tried the experiment, and in 1927 he filed a patent for making helium this way.

The patent was rejected on the grounds that it was too sketchy to be comprehensible. The work foundered, and it was soon discovered that the helium was not produced by fusion at all. It was being absorbed from the atmosphere into the glass walls of the vessels used for the experiments. In 1930, no lesser authorities than James Chadwick and Ernest Rutherford dismissed the claims of hydrogen fusion, saying: 'The presence of an element has been mistaken for its creation.'

No doubt Chadwick and Rutherford would have been quick to pronounce similarly on the experiments of Pons and

Fleischmann, who announced on 23 March 1989 that they had observed 'sustained nuclear fusion' from the electrolysis of heavy water using palladium electrodes. Deuterium is absorbed by palladium in the same way as hydrogen, but its fusion into helium does not require such extreme conditions (see page 136). All the same, these conditions have long proved impossible to sustain in physicists' attempts to harness nuclear fusion for energy generation. Now two chemists were claiming that these massively expensive fusion projects could be abandoned; all you needed was a test tube and two strips of palladium.

Pons and Fleishmann and other groups speculated that the fusion might be happening in tiny cracks in the metal where the pressure on the absorbed deuterium would be greatest. But physicists calculated that these conditions should be nowhere near extreme enough to produce fusion. Despite several announcements in the ensuing months of successful 'cold fusion' in other laboratories, no one was able to demonstrate reproducible and sustained generation of 'excess energy' from the electrolysis cells due to putative fusion reactions. The initial claim of Pons and Fleischmann was made largely on the grounds that they had measured such an energy excess, but some researchers pointed out that, if this energy release was really due to deuterium fusion, it should also have released a lethal dose of neutron radiation. Moreover, concentrating this much hydrogen raised the prospect of a purely chemical explosion—indeed, Pons and Fleischmann did report a virtual 'meltdown' of their electrolysis experiment on one occasion.

By the end of 1989 cold fusion was discredited by all but a minority of true believers (who were still pursuing it over ten years later), and scientists emerged with embarrassment, indignation—and a renewed appreciation of the unique properties of palladium.

Earths rare and coloured

When the Swedish chemist Carl Gustav Mosander discovered lanthanum in 1839, he had no idea what he had started. He extracted it as its oxide—an 'earth'—from cerium nitrate. Mosander's colleague Berzelius suggested the name, from the Greek *lanthanein*: to lie hidden.

But he did not formally announce the new element for two years, because he suspected that it was not wholly pure. In 1841 he revealed that it was mixed with another 'earth', containing an element that he called didymium (from the Greek *didumos*, 'twin').

Yet that was not the end of it. Other chemists suspected that didymium too was not a pure element, but a mixture. Separating its components chemically was very difficult, as they seemed to behave almost identically. But their presence was revealed by inspecting the 'bar code' of elemental emission lines in the glow produced when the material was heated.

In 1879 Paul-Émile Lecoq, gallium's discoverer, announced that there was another element contaminating didymium, which he called samarium. A year later Charles

Galissard de Marignac in Geneva found a further 'earth' in this substance, which Lecoq isolated in 1886 and called gadolinium. Didymium itself, meanwhile, was revealed as a phantom, a mixture of two new elements that Karl Auer in Austria discovered in 1885 and called neodymium ('new didymium') and praseodymium ('green didymium'). Just how many of these 'earth' elements were there, after all?

There are in fact fourteen, and they became known as the rare earths—a misnomer, for some are not particularly rare at all, and they are metals, not 'earths'. A better name is the lanthanides, since they all follow after lanthanum in the Periodic Table.* They form an entirely new group, which cannot economically be fitted into Mendeleyev's scheme and is usually depicted as floating freely below it. The lanthanides are, broadly speaking, all rather similar in their chemical behaviour, which is why they were so hard to separate. They are found in minerals such as monazite and bastnäsite, the main sources of which are in China and the USA.

In 1901 Eugène-Anatole Demarçay in Paris showed that the samples of samarium and gadolinium produced until that time harboured yet another rare-earth element, which he named generously after all of Europe: europium. This element is in fact one of the most naturally abundant of the group: the Earth's crust contains twice as much europium as tin. It is harvested today largely for a very special and useful property: its emission of very 'pure' red and blue light.

---

* The lanthanides proper do not include lutetium, element 71, although this is considered a rare-earth element.

Europium, like all other rare-earth elements, generally forms compounds in which the metal atoms lose three electrons to become ions with three positive charges. This type of europium ion can emit light in the richest red part of the visible spectrum, when suitably stimulated by an energy source. But, unlike the other lanthanides (with the exception of samarium), europium also readily forms an ion that is only doubly charged—deficient in two electrons—which emits rich blue light instead.

Both types of europium ion are incorporated into the phosphors used in colour television screens and computer monitors. Phosphors are substances that emit light when struck by a beam of electrons. The electron beam stimulates electrons in the atomic constituents of the phosphor into states of greater energy, from which they decay back to their initial state by radiating away the excess energy as visible light.

All colours can in principle be created by mixing light of the three primary colours—roughly speaking, red, blue, and green.* In a TV screen the light is mixed by placing three tiny dots of primary-colour phosphors so close together that your eye cannot distinguish them from normal viewing distances.

There are several substances that produce light of these

---

* These are not the same as the three primaries familiar to painters: red, blue, and yellow. This is because mixing lights (additive mixing) is not like mixing pigments (subtractive mixing). Red and green light, for instance, mix to yellow, whereas the corresponding pigments give a dirty brownish colour. And blue and yellow light create not green but white.

three colours when struck by an electron beam. But any old red, blue, and green will not do. The range of colours available from any set of primaries depends on how 'good' a red, blue, and green you start with: if your blue is too pale or greenish, for instance, no amount of colour mixing will give you the deep royal blue of the desert twilight. To obtain good colour pictures on a TV screen, you need phosphors that produce rich, pure primaries. The reds of colour TVs were never very vibrant until, in the early 1960s, manufacturers started to use europium.

Europium satisfies the needs of both red and blue phosphors. Typical materials used for the former are europium yttrium vanadate and yttrium oxysulphide doped with europium. Blue phosphors are made from europium-doped strontium aluminate. The green phosphor in TV screens is typically zinc cadmium sulphide, which cannot produce strongly saturated greens. This means that there are some colours in verdant nature that your TV screen still cannot match; it can deliver only a poor approximation.

Green light is also emitted from some lanthanide elements: lanthanum, cerium, and terbium. A mixture of lanthanide compounds provides all three primary colours in a type of low-energy light bulb called a trichromatic fluorescent bulb. This device contains rare-earth phosphor materials that glow in response not to an electron beam but to ultraviolet light from a mercury arc: an electric discharge sent through mercury vapour. In effect, the phosphors downgrade the high-energy ultraviolet light into visible light. The red component is again provided by a phosphor containing a mixture of europium and yttrium, and the blue by (doubly charged) europium alone. The mixture of red, green, and blue light looks white. Trichromatic bulbs last for much longer than normal incandescent bulbs (which rely on a white-hot filament), and they use a fraction of the power.

The lazy gas

Mendeleyev's Periodic Table of 1869 not only had gaps; it was missing an entire group of elements. It was hardly surprising that no one had found them, because they do not react with other elements to form compounds. They are the noble gases (also called the inert or rare gases) and they comprise the last group of the full table.

The lightest noble gas, helium, had in fact been discovered in 1868—but only on the sun (see page 91). So little was known about it that Mendeleyev could see no way to include it. Helium was not found on Earth until 1895, when William Ramsay and Morris Travers in London isolated it from uranium minerals. Two Swedish chemists in Uppsala found it in much the same source at the same time.

Ramsay had already found another noble gas a year earlier. This one is by no means rare: there are about sixty-six trillion tonnes of it in the atmosphere. It was christened argon, after the Greek *argos*, 'lazy'—because it did nothing.

Air contains almost 1 per cent argon. That is enough to have been noticed by the careful 'pneumatick' chemists of the eighteenth century: Henry Cavendish noted in 1785 that

1 per cent of air seemed to resist any tendency to combine with other elements. But he did not pursue this observation, and it was forgotten.

In the early 1890s, the British physicist Lord Rayleigh found that nitrogen obtained by two different means seemed to have a different density: that extracted from air was very slightly denser than that made by decomposing ammonia (a compound of nitrogen and hydrogen). He and Ramsay investigated both forms of nitrogen, and Ramsay found that atmospheric nitrogen had an inert component that he was finally able to separate. They were able to collect only tiny amounts. Rayleigh lamented in 1894 that 'The new gas has been leading me a life. I had only about a quarter of a thimbleful. I now have a more decent quantity but it has cost about a thousand times its weight in gold.'

Ramsay was nevertheless able to verify its status as a new element by the then-familiar method of observing the spectral lines it emitted. Rayleigh and Ramsay announced the discovery of argon in 1894. Ramsay realized that argon and helium might be members of a hitherto unsuspected new group in the Periodic Table. He and Travers made careful studies of liquid argon, and in 1898 the pair found that it was mixed with tiny quantities of three other noble gases: neon ('new'), krypton ('hidden'), and xenon ('stranger'). This work earned Ramsay the 1904 Nobel Prize in chemistry. (Rayleigh was awarded the physics prize that same year.)

There is one more noble gas in the group: radon, the heaviest of them, which was discovered in 1900 by the German Friedrich Ernst Dorn as a product of the radioactive decay of radium. Ramsay made enough of it to measure its properties in 1908.

Argon is now available in far greater quantities than Ramsay and Rayleigh could glean—over 750,000 tonnes a year are extracted from liquefied air. At first sight it does not exactly strike one as a useful element—for who wants to employ a lazy worker that does nothing? But this inertness is argon's strength. It is the perfect gas if you simply want to bolster an empty space against the mighty push of atmospheric pressure: it is a kind of 'vacuum with pressure'. Thus argon is used to fill tungsten filament bulbs and fluorescent tubes: no matter how hot the filament gets, argon will not react with it. Argon is also used in state-of-the-art double glazing. A vacuum between the two panes of glass would minimize heat conduction across it, but the panes would be pushed together by air pressure. Argon is a poorer conductor of heat than air, and so using it to maintain the pressure between the panes results in less heat loss than air-filled double glazing.

Argon is also an ideal 'carrier gas', a propellant with no propensity to react. A jet of argon is used to stir oxygen into molten iron during steel making: the oxygen reacts with carbon, adjusting its content in the metal. Argon is used to propel sprays of small particles in various technological processes. One can be confident that argon is not going to react in such mixtures: the first chemical compound of argon was made only in 2000, and it is an exotic substance so tenuously bound that it falls apart unless cooled below minus 246 °C.

Although they all contain the same three subatomic constituents, the elements provide a fantastically varied palette for technologists. Their diversity is one of nature's wonders: it is deeply strange, however rationalizable, that yellow sulphur sits between flaming phosphorus and acrid green chlorine. No cook could ever match the natural genius that brews such riches from simple ingredients. And, though the exciting days of element discovery are over (save for those unwieldy superheavies that humankind can make, a few fleeting atoms at a time), the possibilities that the elements offer in combination have by no means been exhausted. Indeed, *that* journey may still be only just beginning.

Notes

1. Aristotle's Quartet: The Elements in Antiquity

13 'The four elements are not a conception'. N. Frye, introduction to
G. Bachelard, *The Psychoanalysis of Fire* (London: Quartet Books,
1987), p. ix.

'I believe it is possible'. G. Bachelard, *Water and Dreams* (Dallas:
Pegasus Foundation, 1983), 3.

14 'the region we call home'. Ibid. 8.

23 'Out of some bodies'. R. Boyle, *The Sceptical Chymist* (1661),
quoted in W. H. Brock, *The Fontana History of Chemistry* (London:
Fontana, 1992), 57.

'certain primitive and simple'. R. Boyle (1661). *The Sceptical
Chymist* (1661), quoted in H. Boynton (ed.), *The Beginnings of
Modern Science* (Roslyn, NY: Walter J. Black Inc., 1948), 254.

2. Revolution: How Oxygen Changed the World

27 *Oxygen*, by C. Djerassi and R. Hoffmann, is published by Wiley-
VCH, Weinheim, 2001.

30 'We have not pretended'. A. L. Lavoisier, *Elements of Chemistry*
(1789), trans. R. Kerr (1790), quoted in R. Boynton (ed.), *The
Beginnings of Modern Science* (Roslyn, NY: Walter J. Black Inc.,
1948), 268–9.

41 'Chemists have made phlogiston a vague principle'. A. L. Lavoisier
(1785), quoted in W. H. Brock, *The Fontana History of Chemistry*
(London: Fontana, 1992), 111–12.

'The same body can pass': Lavoisier (1773), quoted in Brock, *The
Fontana History of Chemistry*, 98.

42 'It is not enough for a substance to be simple': C. Coulston
Gillispie (ed.), *Dictionary of Scientific Biography* (New York:

Scribner's, 1976), viii. 82; quoted in C. Cobb and H. Goldwhite, *Creations of Fire* (New York: Plenum, 1995), 161.

43 The detection of light from an extrasolar planet was reported by A. C. Cameron, K. Horne, A. Penny, and D. James, 'Probable Detection of Starlight Reflected from the Giant Planet Orbiting τ Boötis', *Nature*, 402 (1999), 751.

3. Gold: The Glorious and Accursed Element

51 'He breaks all law'. Virgil, *Aeneid*, iii. l. 55, quoted in G. Agricola, *De re metallica* (1556), trans. H. C. Hoover and L. H. Hoover (New York: Dover, 1950), 16.

'This is indeed the Golden Age'. Quoted in ibid. 10.

52 'It is almost our daily experience'. Ibid. 10.

53 'I have come to take from them their gold'. Pizarro, quoted in L. B. Wright, *Gold, Glory, and the Gospel: The Adventurous Lives and Times of the Renaissance Explorers* (New York: Atheneum, 1970), 229.

'Gold is the universal prize'. J. Bronowski, *The Ascent of Man* (London: Book Club Associates, 1973), 134.

58 'Out of these laborious mines'. Quoted by Hoover and Hoover in ibid. 279 n. 8.

59 'Gold is found in the world'. Pliny, *Natural History*, xxxiii. 21.

60 'The Colchians placed the skins of animals'. Agricola, *De re metallica*, 330.

63 Extracting gold into plant tissues is described by C. W. N. Anderson, R. R. Brooks, R. B. Stewart, and R. Simcock, 'Harvesting a Crop of Gold in Plants', *Nature*, 395 (1998), 553–4.

67 'Dost thou not know the value of money'. Horace, *Satires*. i, l. 73.

68 'When ingenious and clever men'. Agricola, *De re metallica*, 17.

The economic history of gold is engagingly told in P. L. Bernstein, *The Power of Gold* (New York: Wiley, 2000).

70 'We have gold'. Quoted in ibid. 346.

71 'Currencies'. R. Mundell, quoted in *Wall Street Journal*, 10 Dec. 1999, Op-Ed page.

73 'As to the True Man'. Quoted in J. C. Cooper (1990), *Chinese Alchemy* (New York: Sterling Publishing Co., 1990), 66.

76 The chemical state of gold in gold-ruby glass was deduced only very recently: see F. E. Wagner *et al.*, 'Before Striking Gold in Gold-Ruby Glass', *Nature*, 407 (2000), 691–2.

79 The explanation for the inertness of gold is given in B. Hammer and J. K. Nørskov, 'Why Gold is the Noblest of all the Metals', *Nature*, 376 (1995), 238–40.

4. The Eightfold Path: Organizing the Elements

87 'encouraged people to acquire a faith'. W. H. Brock, *The Fontana History of Chemistry* (London: Fontana, 1992), 139–40.

88 'Berzelius's symbols are horrifying.' Quoted in ibid. 139.

92 The concept of *prote hyle* and its relation to early ideas about stellar evolution are discussed in S. F. Mason, *Chemical Evolution* (Oxford: Clarendon Press, 1992).

93 'It was quite the most incredible event'. Quoted in G. K. T. Conn and H. D. Turner, *The Evolution of the Nuclear Atom* (London: Iliffe Books, 1965), 136.

98 'he is a nice sort of fellow'. Quoted in B. Jaffe, *Crucibles: The Story of Chemistry* (New York: Dover, 1976), 151.
'I saw in a dream'. Quoted in P. Strathern, *Mendeleyev's Dream* (London: Penguin, 2000), 286.

5. The Atom Factories: Making New Elements

113 One of the best accounts of the early development of atomic and nuclear chemistry is R. Rhodes, *The Making of the Atom Bomb* (New York: Simon & Schuster, 1986).

119 'some fool in a laboratory'. A. S. Eve, *Rutherford* (London: Macmillan, 1939), 102.
'The man who put his hand on the lever'. F. Soddy, *Atomic Transmutation* (New World, 1953), 95.

126 'Your results are very startling'. L. Meitner, letter to O. Hahn, 21 Dec. 1938, reproduced in J. Lemmerich (ed.), *Die Geschichte der*

Entdeckung der Kernspaltung: Austellungskatalog (Technische Universität Berlin, Universitätsbibliothek, 1988), 176. See also R. Lewin Sime, *Lise Meitner: A Life in Physics* (Berkeley and Los Angeles: University of California Press, 1996), 235.

127 'when fission was discovered'. Quoted in C. Weiner (ed.), *Exploring the History of Nuclear Physics*, AIP Conference Proceedings No. 7 (American Institute of Physics, 1972), 90.

132 'To change the hydrogen in a glass'. F. Aston, 'Forty Years of Atomic Theory', in J. Needham and W. Pagel (eds.), *Background to Modern Science* (London: Macmillan, 1938), 108.
'We can only hope'. Ibid. 114.

134 'first step along the way'. I. McEwan, *Enduring Love* (London: Vintage, 1998), 3.

144 A brief and good account of the manufacture of superheavy elements, and the search for the island of stability, is given by R. Stone, *Science*, 278 (1997), 571, and *Science*, 283 (1999), 474. The topic is discussed in more detail in G. T. Seaborg and W. D. Loveland, 'The search for new elements', in N. Hall (ed.), *The New Chemistry* (Cambridge: Cambridge University Press, 2000), 1.

146 The chemical properties of seaborgium are described in M. Schädel *et al.*, 'Chemical Properties of Element 106 (Seaborgium)', *Nature*, 388 (1997), 55.

6. The Chemical Brothers: Why Isotopes are Useful

149 The story of the discovery of Oetzi the 'iceman' is told in B. Fowler, *Iceman* (New York: Random House, 2000).

156 The radiocarbon dating of the Shroud of Turin is described in P. E. Damon *et al.*, 'Radiocarbon Dating of the Shroud of Turin', *Nature*, 337 (1989), 611.

160 The evidence for the oldest zircons, and their interaction with water, is described in S. A. Wilde, J. W. Valley, W. H. Peck, and C. M. Graham, 'Evidence from Detrital Zircons for the Existence of Continental Crust and Oceans on the Earth 4.4 Gyr Ago', *Nature*, 409 (2001), 175, and S. J. Mojzsis, T. M. Harrison, and

R. T. Pidgeon, 'Oxygen-Isotope Evidence from Ancient Zircons for Liquid Water at the Earth's Surface 4,300 Myr Ago', *Nature*, 409 (2001), 178.

161 The uranium dating of a star is reported in R. Cayrel *et al.*, 'Measurement of Stellar Age from Uranium Decay', *Nature*, 409 (2001), 691.

164 The new understanding of global climate change that emerged from the CLIMAP studies of deep-sea sediment cores is described in J. Imbrie and K. P. Imbrie, *Ice Ages* (Cambridge, Mass.: Harvard University Press, 1986).

165 Fig. 15*a* comes from N. J. Shackleton, A. Berger, and W. R. Peltier, 'An Alternative Astronomical Calibration of the Lower Pleistocene Timescale Based on ODP Site 677', *Transactions of the Royal Society of Edinburgh Earth Sciences*, 81 (1990), 251. Fig. 15*b* comes from J. Jouzel *et al.*, 'Extending the Vostok Ice-Core Record of Palaeoclimate to the Penultimate Glacial Period', *Nature*, 364 (1993), 407.

7. For All Practical Purposes: Technologies of the Elements

187 'The presence of an element'. J. Chadwick, C. D. Ellis, and E. Rutherford, *Radiation from Radioactive Substances* (Cambridge: Cambridge University Press, 1930). The woeful tale of cold fusion, and the precedent in the work of Pareth and Peters, is recounted in F. Close, *Too Hot To Handle* (Princeton: Princeton University Press, 1991).

193 The many industrial uses of rare-earth elements are described in D. Lutz, 'The Quietly Expanding Rare-Earth Market', *The Industrial Physicist*, 2/3 (Sept. 1996), 28.

194 'The new gas has been leading me a life'. B. Jaffe, *Crucibles: The Story of Chemistry* (New York: Dover, 1976), 159.

195 The first compound of argon is reported in L. Khriachtchev, M. Pettersson, N. Runeberg, J. Lundell, and M. Räsänen, 'A Stable Argon Compound', *Nature*, 406 (2000), 874.

Further reading

Atkins, P., *The Periodic Kingdom* (London: Weidenfeld & Nicolson, 1995).

Brock, W. H., *The Fontana History of Chemistry* (London: Fontana, 1992).

Cobb, C., and Goldwhite, H., *Creations of Fire* (New York: Plenum, 1995).

Emsley, J., *Nature's Building Blocks* (Oxford: Oxford University Press, 2001).

Gray, H. B., Simon, J. D., and Trogler, W. C., *Braving the Elements* (Sausalito, Calif.: University Science Books, 1995).

Jaffe, B., *Crucibles: The Story of Chemistry* (New York: Dover, 1976).

Sass, S. L. (1998), *The Substance of Civilization* (New York: Arcade, 1998).

Strathern, P., *Mendeleyev's Dream* (London: Penguin, 2000).

Information about the elements is available online at:
www.webelements.com.

Index